LIFE ASCENDING

The Ten Great Inventions of Evolution

NICK LANE

PROFILE BOOKS

This paperback edition published in 2010

First published in Great Britain in 2009 by
PROFILE BOOKS LTD
3A Exmouth House
Pine Street
London EC1R 0JH
www.profilebooks.com

1 3 5 7 9 10 8 6 4 2

Printed and bound in Great Britain by
Bookmarque Ltd, Croydon, Surrey

Typeset in Fournier by MacGuru Ltd
info@macguru.org.uk

A CIP catalogue record for this book is available from the British Library.

ISBN 978 1 86197 818 9
eISBN 978 1 84765 222 5

LIFE ASCENDING

'Lane is that particularly rare breed: a scientist who can not only offer a bird's-eye view of an entire field but also tell you about his own very interesting ideas.' Carl Zimmer, *Science*

'An original and awe-inspiring account. The first two chapters are the most coherent and convincing summaries of the dawn of life I have ever read … this is an exhilarating tour of the most profound and important ideas in biology. Anyone interested in life should read it. Highly recommended.' *New Scientist*

'An award-winning biochemist reconstructs the history of life in focusing on the ten greatest inventions of evolution … Lane confers vivid and revealing insights into the nature of our own existence on planet Earth.' *Good Book Guide*

'Wonderful … Lane does a masterful job … an elegant, fully satisfying whole.' Starred Review, *Publishers Weekly*

'An excellent book. It is great fun, readable, bubbling over with enthusiasm, and not afraid of controversial, even weird, ideas … Hopeless as a bedside book: you'd never sleep.' Graham Cairns-Smith, *Chemistry World*

'A fascinating look at how scientists have come to understand evolution with an ingenuity rivaling that of nature herself.' *Science News*

'A clever concept is carried through with a clarity and enthusiasm that belies the sophistication of the science …' *Guardian*

'*Life Ascending* really is beautifully written, and Lane has a true flair for scientific story telling … utterly gripping.' *Astrobiology Society of Britain*

'Lane lays out processes of dizzying complexity in smooth, nimble prose.' *Kirkus Reviews*

'A fascinating and beautifully written account of the great mysteries of life – how it arose, how it works, why things die, how consciousness evolved. It's a great read, and provides real insight into current scientific thinking about the big evolutionary puzzles without getting tangled up in difficulties.' Ian Stewart, author of *Professor Stewart's Cabinet of Mathematical Curiosities*

'A vivid picture of how evolution has informed life.' *Globe and Mail* (Canada)

'Terrific explanation of the newest evolutionary findings.' *Sunday Star Times* (New Zealand)

'A writer who is not afraid to think big – and think hard' Frank Wilczek, 2004 Nobel Laureate in Physics

NICK LANE is a biochemist and Provost's Venture Research Fellow in the Department of Genetics, Evolution and Environment at University College London. His previous book, *Power, Sex, Suicide* was short-listed for the *Royal Society Science Book Prize* and *THE Young Academic Author of the Year*, and named a book of the year by the *Economist*. He lives in London. For more information on Dr Lane visit *www.nick-lane.net*

CONTENTS

For my mother and father
Now I am a parent I appreciate all you have done for me more than ever

INTRODUCTION

The Ten Great Inventions of Evolution

Set against the consuming blackness of space, the earth is a beguiling blue-green ball. Barely two dozen people have ever experienced the emotion of seeing our planet from the moon and beyond, yet the fragile beauty of the pictures they sent back home is engraved in the minds of a generation. Nothing compares. Petty human squabbles over borders and oil and creed vanish in the knowledge that this living marble surrounded by infinite emptiness is our shared home, and more, a home we share with, and owe to, the most wonderful inventions of life.

Life itself transformed our planet from the battered and fiery rock that once orbited a young star, to the living beacon that is our world seen from space. Life itself turned our planet blue and green, as tiny photosynthetic bacteria cleansed the oceans of air and sea, and filled them with oxygen. Powered by this new and potent source of energy, life erupted. Flowers bloom and beckon, intricate corals hide darting gold fish, vast monsters lurk in black depths, trees reach for the sky, animals buzz and lumber and see. And in the midst of it all, we are moved by the untold mysteries of this creation, we cosmic assemblies of molecules that feel and think and marvel and wonder at how we came to be here.

For the first time in the history of our planet, we know. This is no certain knowledge, no stone tablet of truth, but the ripening fruits of mankind's greatest quest, to know and understand the living world around us and within us. We have understood in outline since Darwin, of course, whose *Origin of Species* was published 150 years ago. Since Darwin our knowledge of the past has been fleshed out not only with fossils filling in gaps, but with an understanding of the intimate structure of the gene – an understanding that now underpins every stitch in the rich tapestry of life. And yet it is only in the last decades that we have moved from theory and abstract knowledge to a vibrant and detailed picture of life, written in languages that we have only recently begun to translate, and which hold the keys not only to the living world around us but also to the most remote past.

The story that unfurls is more dramatic, more compelling, more intricate than any creation myth. Yet like any creation myth, it is a tale of transformations, of sudden and spectacular changes, eruptions of innovation that transfigured our planet, overwriting past revolutions with new layers of complexity. The tranquil beauty of our planet from space belies the real history of this place, full of strife and ingenuity and change. How ironic that our own petty squabbles reflect our planet's turbulent past, and that we alone, despoilers of the Earth, can rise above it to see the beautiful unity of the whole.

Much of this planetary upheaval was catalysed by two handfuls of evolutionary innovations, inventions that changed the world and ultimately made possible our own lives. I must clarify what I mean by invention, for I don't want to imply a deliberate inventor. The Oxford English Dictionary defines an invention as: 'The original contrivance or production of a new method or means of doing something, previously unknown; origination, introduction.' Evolution has no foresight, and does not plan for the future. There is no inventor, no intelligent design. Nonetheless, natural selection subjects all traits to the most exacting tests, and the best designs win out. It is a natural laboratory that belittles the human theatre, scrutinising trillions of tiny differences simultaneously, each and every generation. Design is all around us, the product of blind but ingenious processes. Evolutionists often talk informally of inventions, and there is no better word to convey the astonishing creativity of nature. To gain an insight into how all this came about is the shared goal of scientists, whatever their religious beliefs, along with anyone else who cares about how we came to be here.

This book is about the greatest inventions of evolution, how each one transformed the living world, and how we humans have learned to read this past with an ingenuity that rivals nature herself. It is a celebration of life's marvellous inventiveness, and of our own. It is, indeed, the long story of how we came to be here – the milestones along the epic journey from the origin of life to our own lives and deaths. It is a book grand in scope. We shall span the length and breadth of life, from its very origins in deep-sea vents to human consciousness, from tiny bacteria to giant dinosaurs. We shall span the sciences, from geology and chemistry to neuroimaging, from quantum physics to planetary science. And we shall span the range of human achievement, from the most celebrated scientists in history to researchers as yet little known, if destined one day, perhaps, to be famous.

My list of inventions is subjective, of course, and could have been different; but I did apply four criteria which I think restrict the choice considerably to a few seminal events in life's history.

The first criterion is that the invention had to revolutionise the living world, and so the planet as a whole. I mentioned photosynthesis already, which turned the Earth into the supercharged, oxygen-rich planet we know (without which animals are impossible). Other changes are less obvious, but almost equally pervasive. Two inventions with the most widespread consequences are movement, which allowed animals to range around in search of food, and sight, which transformed the character and behaviour of all living organisms. It may well be that the swift evolution of eyes, some 540 million years ago, contributed in no small measure to the sudden appearance of proper animals in the fossil record, known as the Cambrian explosion. I discuss the Earth-moving consequences of each invention in the introductions to individual chapters.

My second criterion is that the invention had to be of surpassing importance today. The best examples are sex and death. Sex has been described as the ultimate existential absurdity, and that is to ignore the Kama Sutra's worth of contorted mental postures, from angst to ecstasy, and focus only on the peculiar mechanics of sex between cells. Why so many creatures, even plants, indulge in sex when they could just quietly clone copies of themselves is a conundrum that we are now very close to answering. But if sex is the ultimate existential absurdity, then death must be the ultimate non-existential absurdity. Why do we grow old and die, suffering along the way the most harrowing and dreadful diseases? This most modern preoccupation is not dictated by thermodynamics, the laws of mounting chaos and corruption, for not all living things age, and even those that do can flip a switch and stop. We shall see that evolution has extended the lifespan of animals by an order of magnitude, time and again. The anti-ageing pill should not be a myth.

The third criterion is that each invention had to be a direct outcome of evolution by natural selection, rather than, for example, cultural evolution. I am a biochemist, and I have nothing original to say about language or society. And yet the substrate for all we have achieved, all that is human, is consciousness. It is hard to picture any form of shared language or society that is not underpinned by shared values, perceptions or feelings, wordless feelings like love, happiness, sadness, fear, longing, hope, belief. If the human mind evolved, we must explain how nerves firing in the brain can give rise to the sense of immaterial spirit, to the subjective intensity of feelings. For me, this is a biological problem, if still a vexed one, as I endeavour to justify in Chapter 9. So consciousness is 'in' as one of the great inventions; language and society are out, as more the products of cultural evolution.

My final criterion is that the invention had to be iconic in some way. The supposed perfection of the eye is perhaps the archetypal challenge, dating back to Darwin and before. Since then the eye has been addressed many times, in many ways, but the explosion of genetic insights in the last decade offers a new resolution, an unexpected ancestry. The spiralling double helix of DNA is the greatest icon of our information age. The origin of complex cells (the 'eukaryotic' cell) is another iconic subject, albeit better known among scientists than the lay public. This milestone has been one of the most hotly contested matters among evolutionists over the last four decades, and is crucially important to the question of how widespread complex life might be across the universe. Each chapter deals in its way with iconic issues such as these. At the outset, I discussed my list with a friend, who proposed 'the gut' as emblematic of animals, in place of movement. The idea falters in its status as an icon: to my mind at least the power of muscle is iconic – think only of the glories of flight – the gut, without powered movement, is but a sea squirt, a swaying pillar of intestines tied to a rock. Not iconic.

Beyond these more formal criteria, each invention had to catch my own imagination. These are the inventions that I, as a passionately curious human being, most wanted to understand myself. Some I have written about before, and wanted to address again in a broader setting; others, like DNA, exert a kind of fatal attraction for all inquisitive minds. The unravelling of clues buried deep in its structure is one of the great scientific detective stories of the last half-century, and yet somehow remains little known even among scientists. I can only hope I have succeeded in conveying some of my own thrill in the chase. Hot blood is another example, an area of furious controversy, for there is still little consensus about whether the dinosaurs were active hot-blooded killers, or slothsome giant lizards, whether the hot-blooded birds evolved directly from close cousins of *T. rex*, or had nothing to do with dinosaurs. What better chance to review the evidence myself!

So we have a list. We start with the origin of life itself, and end with our own deaths and prospects for immortality, by way of such pinnacles as DNA, photosynthesis, complex cells, sex, movement, sight, hot blood and consciousness.

But before we start, I must say a few words about the leitmotiv of this introduction: the new 'languages' that afford such insights into the depths of evolutionary history. Until recently, there have been two broad paths into the past: fossils and genes. Both have enormous power to breathe life into the past, but each has its flaws. The supposed 'gaps' in the fossil record are over-sung, and many have been laboriously filled in over the 150 years since Darwin worried about them. The trouble is that fossils, through the very conditions that favour their preservation, cannot and do not hold an undistorted mirror to the past. The fact that we can glean so much from them is remarkable. Likewise, comparing the details of gene sequences enables us to build genealogical trees, which show precisely how we are related to other organisms. Unfortunately, genes ultimately diverge to the point that they no longer have anything in common: beyond a certain point, the past, as read by genes, becomes garbled. But there are powerful methods that go beyond genes and fossils, far back into the deepest past, and this book is in part a celebration of their acuity.

Let me give a single example, one of my own favourites, which never found the opportunity for a mention in the book proper. It concerns a protein (a catalyst, or enzyme, called citrate synthase) which is so central to life that it is found in all living organisms, from bacteria to man. This enzyme has been compared in two different species of bacteria, one living in superhot hydrothermal vents, the other in the frozen Antarctic. The gene sequences that encode these enzymes are different; they have diverged to the point that they are now quite distinct. We know that they *did* diverge from a common ancestor, for we see a spectrum of intermediates in bacteria living in more temperate conditions. But from the gene sequences alone there is little more we can say. They diverged, surely because their living conditions are so different, but this is abstract theoretical knowledge, dry and two-dimensional.

But now look at the molecular structure of these two enzymes, pierced by an intense beam of X-rays and deciphered through the wonderful advances in crystallography. The two structures are superimposable, so similar to each other that each fold and crevice, each niche and protrusion, is faithfully replicated in the other, in all three dimensions. An untutored eye could not distinguish between them. In other words, despite a large number of building blocks being replaced over time, the overall shape and structure of the molecule – and thus its function – has been preserved throughout evolution, as if it were a cathedral built in stone, and rebuilt from within using bricks, without losing its grand architecture. And then came another revelation. Which building blocks got switched and why? In the superhot vent bacteria, the enzyme is as rigid as possible. The building blocks bind tightly to each other, through internal bonds that work like cement, retaining the structure despite the buffeting of energy from the boiling vents. It is a cathedral built to withstand perpetual earthquakes. In the ice, the picture is reversed. Now the building blocks are flexible, allowing movement despite the frost. It's as if the cathedral were rebuilt with ball-bearings, rather than bricks. Compare their activity at 6°C, and the frosty enzyme is twenty-nine times as fast; but try at 100°C, and it falls to pieces.

The picture that emerges is colourful and three-dimensional. The changes in gene sequence now have meaning: they preserve the structure of the enzyme and its function, despite the need to operate under totally different conditions. We can now see what actually happened over evolution, and why. It is no longer merely intimation, but real insight.

Similarly vivid insights into what actually happened can be gained from other clever tools now available. Comparative genomics, for one, allows us to compare not just genes, but full genomes, thousands of genes at once, in hundreds of different species. Again, this has only been possible in the last few years, as whole genome sequences have proliferated. Then proteomics allows us to capture the spectrum of proteins working within a cell at any one time, and to grasp how this spectrum is controlled by a small number of regulatory genes that have been preserved down the aeons of evolution. Computational biology enables us to identify particular shapes and structures, motifs, which persist in proteins despite changes in genes. Isotopic analyses of rocks or fossils allow us to reconstruct past changes in the atmosphere and climate. Imaging techniques let us to see the function of neurons in the brain while we think, or to reconstruct the three-dimensional structure of microscopic fossils embedded in rocks without disturbing them. And so on.

None of these techniques is new. What *is* new is their sophistication, speed and availability. Like the Human Genome Project, which accelerated to a crescendo well ahead of schedule, the pace at which data are accumulating is dizzying. Much of this information is written not in the classical tongues of population genetics and palaeontology, but in the language of molecules, the level at which change actually occurs in nature. With these new techniques, a new breed of evolutionist is emerging, able to capture the workings of evolution in real time. The picture so painted is breathtaking in its wealth of detail and its compass, ranging from the subatomic to the planetary scale. And that is why I said that, for the first time in history, we know. Much of our growing body of knowledge is provisional, to be sure, but it is vibrant and meaningful. It is a joy to be alive at this time, when we know so much, and yet can still look forward to so much more.

THE ORIGIN OF LIFE

From Out the Turning Globe

Night followed day in swift succession. On earth at that time a day lasted for only five or six hours. The planet spun madly on its axis. The moon hung heavy and threatening in the sky, far closer, and so looking much bigger, than today. Stars rarely shone, for the atmosphere was full of smog and dust, but spectacular shooting stars regularly threaded the night sky. The sun, when it could be seen at all through the dull red smog, was watery and weak, lacking the vigour of its prime. Humans could not survive here. Our eyes would not bulge and burst, as they may on Mars; but our lungs could find no breath of oxygen. We'd fight for a desperate minute, and asphyxiate.

The earth was named badly. 'Sea' would have been better. Even today, oceans cover two-thirds of our planet, dominating views from space. Back then, the earth was virtually all water, with a few small volcanic islands poking through the turbulent waves. In thrall to that looming moon, the tides were colossal, ranging perhaps hundreds of feet. Impacts of asteroids and comets were less common than they had been earlier, when the largest of them flung off the moon; but even in this period of relative tranquillity, the oceans regularly boiled and churned. From underneath, too, they seethed. The crust was riddled with cracks, magma welled and coiled, and volcanoes made the underworld a constant presence. It was a world out of equilibrium, a world of restless activity, a feverish infant of a planet.

It was a world on which life emerged, 3,800 million years ago, perhaps animated by something of the restlessness of the planet itself. We know because a few grains of rock from that bygone age have survived the restless aeons to this very day. Inside them are trapped the tiniest specks of carbon, which bear in their atomic composition the nearly unmistakable imprint of life itself. If that seems a flimsy pretext for a monumental claim, perhaps it is; there isn't a full consensus among experts. But strip away a few more skins from the onion of time and, by 3,400 million years ago, the signs of life are unequivocal. The world was heaving with bacteria then, bacteria that left their mark not just in carbon signatures but in microfossils of many diverse forms and in those domed cathedrals of bacterial life, the metre-high stromatolites. Bacteria dominated our planet for another 2,500 million years before the first truly complex organisms appeared in the fossil record. And some say they still do, for the gloss of plants and animals doesn't match the bacteria for biomass.

What was it about the early earth that first breathed life into inorganic elements? Are we unique, or exceedingly rare, or was our planet but one in a million billion hatcheries scattered across the universe? According to the anthropic principle it scarcely matters. If the probability of life in the universe is one in a million billion, then in a million billion planets there is a chance approaching 1 that life should emerge somewhere. And because we find ourselves on a living planet, obviously we must live on that one. However exceedingly rare life might be, in an infinite universe there is always a probability of life emerging on one planet, and we must live on that planet.

If you find overly clever statistics unsatisfying, as I do, here is another unsatisfying answer, put forward by no lesser statesmen of science than Fred Hoyle and later Francis Crick. Life started somewhere else and 'infected' our planet, either by chance or by the machinations of some god-like extraterrestrial intelligence. Perhaps it did – who would go to the stake to say that it didn't? – but most scientists would back away from such reasoning, with good reason. It is tantamount to an assertion that science cannot answer the question, before we've even bothered to look into whether science can, in fact, answer it. The usual reason given for seeking salvation elsewhere in the universe is time: there has not been enough time, on earth, for the stupefying complexity of life to evolve.

But who says? The Nobel laureate Christian de Duve, equally eminent, argues altogether more thrillingly that the determinism of chemistry means that life had to emerge quickly. In essence, he says, chemical reactions must happen rapidly or not at all; if any reaction takes a millennium to complete, then the chances are that all the reactants will simply dissipate or break down in the meantime, unless they are continually replenished by other, faster, reactions. The origin of life was certainly a matter of chemistry, so the same logic applies: the basic reactions of life must have taken place spontaneously and quickly. So life, for de Duve, is far more likely to evolve in 10,000 years than 10 billion.

We can never know how life really started on earth. Even if we succeed in producing bacteria or bugs that crawl out from swirling chemicals in a test tube, we will never know if that is how life actually started on our planet, merely that such things are possible, and perhaps more likely than we once thought. But science is not about exceptions, it's about rules; and the rules that govern the emergence of life on our own planet should apply throughout the universe. The quest for the origin of life is not an attempt to reconstruct what happened at 6.30 a.m. on Thursday morning in the year 3,851 million BC, but for the general rules that must govern the emergence of any life, anywhere in the universe, and especially on our planet, the only example we know. While the story we'll trace is almost certainly not correct in every particular, it is, I think, broadly plausible. I want to show that the origin of life is not the great mystery it is sometimes made out to be, but that life emerges, perhaps almost inevitably, from the turning of our globe.

Science is not just about rules, of course; it's also about the experiments that elucidate the rules. Our story begins in 1953, then, an *annus mirabilis* marked by the coronation of Queen Elizabeth II, the ascent of Everest, the death of Stalin, the elucidation of DNA, and, not least, the Miller–Urey experiment, the symbolic origin of origin of life research. Stanley Miller was at that time a headstrong doctoral student in the lab of Nobel laureate Harold Urey; he died perhaps a touch embittered in 2007, still fighting for views that he had upheld doughtily for half a century. But whatever the fate of his own particular ideas,

Miller's true legacy was the field that he founded on his remarkable experiments, the results of which retain the power to amaze even today.

Miller filled a large glass flask with water and a mixture of gases, to simulate what he took to be the primordial composition of the earth's atmosphere. The gases he chose were believed (from spectroscopy) to make up the atmosphere of Jupiter, and were reasonably assumed to have been plentiful on the young earth too, to wit, ammonia, methane and hydrogen. Through this mixture, Miller passed electric sparks to simulate lightning, and waited. After a few days, a few weeks, a few months, he took samples and analysed them to determine exactly what he was cooking. And his findings exceeded even his own wildest imaginings.

He was cooking a primordial soup, a near-mythical mix of organic molecules, including a few amino acids, the building blocks of proteins, and probably the most symbolic molecules of life, certainly at that time, before DNA achieved fame. Even more strikingly, the amino acids that actually formed in Miller's soup tended to be the same as those used by life, rather than others drawn randomly from a large reservoir of potential structures. In other words, Miller electrified a simple mixture of gases, and the basic building blocks of life all congealed out of the mix. It was as if they were waiting to be bidden into existence. Suddenly the origin of life looked easy. The idea must have captured something of the spirit of the age, for the story made the cover of *Time* magazine – an unprecedented splash of publicity for a scientific experiment.

Over time, though, the idea of a primordial soup fell out of favour. Its fortunes hit a nadir when analyses of ancient rocks made it plain that the earth had never been rich in methane, ammonia and hydrogen, at least not after the great asteroid bombardment that blasted off the moon. That colossal bombardment shredded the first atmosphere of our planet, sweeping it away into space. More realistic simulations of the primordial atmosphere proved disappointing. Try passing electric discharges through a mixture of carbon dioxide and nitrogen, with trace levels of methane and other gases, and your yield of organic molecules drops dismally. Scarcely an amino acid in sight. The primordial soup became little more than a curiosity, if still a fine demonstration that organic molecules *could* be made by simple means in the lab.

The soup was rescued by the detection of abundant organic molecules in space, most notably on comets and meteorites. A few of these seemed to be composed almost entirely of dirty ice and organic molecules, and served up a remarkably similar array of amino acids to those formed by electrifying gases. Beyond the surprising fact of their existence, it was beginning to look as if there was something specially favoured about the molecules of life – a small subset of the vast library of all possible organic molecules. The great asteroid bombardment now took on quite a different face: no longer merely destructive, the pounding became the ultimate source of all the water and organic molecules needed for life to get going. The soup was not indigenous to Earth but delivered from outer space. And although most organic molecules would have been frazzled on impact, calculations suggested that enough could have survived to stock a soup.

Even if not quite the seeding of life from space advocated by cosmologist Fred Hoyle, the idea nevertheless tied the origins of life, or at least the soup, to the fabric of the universe. Life was no longer a lonely exception, but now a magisterial cosmological constant, inevitable as gravity. Needless to say, astrobiologists loved the idea. Many still do. Quite apart from being a pleasing idea, it gave them job security.

The soup was also pleasingly palatable with molecular genetics, and especially the idea that life is all about replicators, in particular genes, made of DNA or RNA, which can copy themselves with fidelity and pass on to the next generation (more on this in the next chapter). It is certainly true that natural selection can't work without some sort of replicator; and it is equally true that life can *only* evolve complexity through the auspices of natural selection. For many molecular biologists, then, the origin of life *is* the origin of replication. And a soup fits that idea nicely, as it seems to provide all the constituents needed for competing replicators to grow and evolve. In a nice thick soup, the replicators take what they need, forming longer, more complex polymers, and eventually manipulating other molecules into elaborate structures such as proteins and cells. In this view, the soup is an alphabetic sea writhing with letters, just waiting for natural selection to fish them out and turn them into surpassing prose.

For all this, the soup is a pernicious idea. Not pernicious because it's necessarily wrong – there really might have been a primordial soup once upon a time, even if it was a lot more dilute than originally claimed. It is pernicious because the idea of a soup deflected attention away from the true underpinnings of life for decades. Take a large sterilised tin of soup (or peanut butter), and leave it for a few million years. Will life emerge? No. Why not? Because left to themselves the contents will do nothing but break down. If you zap the tin repeatedly, you won't be any better off, for the soup will only break down even faster. A sporadic and massive discharge like lightning might persuade a few sticky molecules to congeal together in clumps, but it is far more likely just to shred them into bits again. Could it create a population of sophisticated replicators in a soup? I doubt it. As the Arkansas traveller has it, 'You can't get there from here.' It's just not thermodynamically reasonable, for the same reasons that a corpse can't be brought back to life by electrocuting it repeatedly.

Thermodynamics is one of those words best avoided in a book with any pretence to be popular, but it's more engaging if seen for what it is: the science of 'desire'. The existence of atoms and molecules is dominated by 'attractions', 'repulsions', 'wants' and 'discharges', to the point that it becomes virtually impossible to write about chemistry without giving in to some sort of randy anthropomorphism. Molecules 'want' to lose or gain electrons; attract opposite charges; repulse similar charges; or cohabit with molecules of similar character. A chemical reaction happens spontaneously if all the molecular partners desire to participate; or they can be pressed to react unwillingly through greater force. And of course some molecules really want to react but find it hard to overcome their innate shyness. A little gentle flirtation might prompt a massive release of lust, a discharge of pure energy. But perhaps I should stop there.

My point is that thermodynamics makes the world go round. If two molecules don't want to react together, then they won't be easily persuaded; if they do want to react they will, even if it takes some time to overcome their shyness. Our lives are driven by wants of this kind. The molecules in food want very much to react with oxygen, but luckily they don't react spontaneously (they're a touch shy), or we'd all go up in flames. But the flame of life, the slow-burning combustion that sustains us all, is a reaction of exactly this type: hydrogen stripped from food reacts with oxygen to release all the energy we need to live.[1] At bottom, *all* life is sustained by a 'main reaction' of a similar type: a chemical reaction that wants to happen, and releases energy that can be used to power all the side-reactions that make up metabolism. All this energy, all our lives, boils down to the juxtaposition of two molecules totally out of equilibrium with each other, hydrogen and oxygen: two opposing bodies that conjoin in blissful molecular union, with a copious discharge of energy, leaving nothing but a small, hot puddle of water.

And that is the problem with the primordial soup: it is thermodynamically flat. Nothing in the soup particularly wants to react, at least not in the way that hydrogen and oxygen want to react. There is no disequilibrium, no driving force to push life up, up, up the very steep energetic hill to the formation of truly complex polymers, such as proteins, lipids, polysaccharides, and most especially RNA and DNA. The idea that replicators like RNA were the first figments of life, predating any thermodynamic driving force, is, in Mike Russell's words, 'like removing the engine from an automobile and expecting the regulating computer to do the driving.' But if not from a soup, where did the engine come from?

The first clue to an answer came in the early 1970s, when rising plumes of warmish water were noticed along the Galapagos Rift, not far from the Galapagos Islands. Appropriately enough, the islands whose richness once seeded the origin of species in Darwin's mind now offered a clue to the origin of life itself.

Little happened for a few years. Then in 1977, eight years after Neil Armstrong set foot on the moon, the US naval submersible *Alvin* descended to the rift, seeking the oceanic hydrothermal vents that presumably gave rise to the warm-water plumes, and duly found them. Yet while their existence hardly came as a surprise, the sheer exuberance of life in the bible-black depths of the rift came as a genuine shock. Here were giant tubeworms, some of them eight feet long, mixed with clams and mussels as big as dinner plates. If giants in the ocean depths were not unusual – just think of the giant squid – their

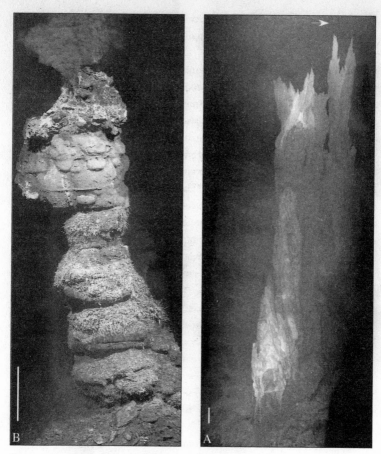

Figure 1.1 A volcanically driven black smoker, venting at 350°C, on the Juan de Fuca Ridge, Northeast Pacific Ocean. The marker is one metre in height.

Figure 1.2 Nature Tower, a 30-metre tall active alkaline vent at Lost City, rising from the serpentine bedrock. The actively venting areas are brighter white. The marker is one metre in height.

sheer abundance was astounding. Population densities in the deep-sea vents rival a rainforest or a coral reef, despite being powered by the exhalations of the vents rather than the sun.

Perhaps most dramatic of all were the vents themselves, which soon acquired the name 'black smokers' (see Fig. 1.1). As it happened, the Galapagos Rift vents were tame affairs compared with some of the other 200 vent fields discovered since, scattered along the ocean ridges of the Pacific, Atlantic and Indian oceans. Tottering black chimneys, some of them tall as skyscrapers, pump billowing black smoke into the oceans above. The smoke is not real smoke, but broiling metal sulphides invading the seawater welling up from the magma furnace below, acidic as vinegar, reaching temperatures of 400°C in the crushing pressure of the ocean deep before precipitating in the cold waters. The chimneys themselves are composed of sulphur minerals like iron pyrites (better known as fool's gold) which settle out from the black smoke, amassing in thick deposits over wide areas. Some chimneys grow at startling rates, as much as thirty centimetres in a day, and can tower to sixty metres before crashing down.

This bizarre and isolated world seemed to be a vision of Hell, and came replete with brimstone and the foul reek of hydrogen sulphide gas emanating from the smokers. Surely only the disturbed mind of Hieronymus Bosch could have imagined the giant tubeworms, lacking either a mouth or an anus, and the eyeless shrimp, swarming in countless multitudes on the ledges beneath the chimneys, grotesque as a plague of locusts. Life in the smokers doesn't just endure these infernal conditions, it can't live without them, it thrives on them. But how?

The answer lies in the disequilibrium. As seawater percolates down to the magma beneath the black smokers, it is superheated and charged with minerals and gases, most notably hydrogen sulphide. Sulphur bacteria can extract hydrogen from this mix and attach it to carbon dioxide to form organic matter. The reaction is the basis of life in the vents, allowing bacteria to flourish with no direct input from the sun. But the conversion of carbon dioxide to organic matter costs energy, and to provide it the sulphur bacteria need oxygen. The reaction of hydrogen sulphide with oxygen releases the energy that powers the vent world and is equivalent to the reaction of hydrogen with oxygen that powers our own lives. The products are water, as before, but then elemental sulphur, the biblical brimstone that gives the sulphur bacteria their name.

It's worth noting that the vent bacteria have no direct use for either the heat or any other aspect of the vent, beyond the hydrogen sulphide emitted.[2] This gas is not inherently rich in energy; it is the reaction with oxygen that provides the energy, and this depends on the *interface* between the vents and the oceans, the juxtaposition of two worlds in dynamic disequilibrium. Only the bacteria living right next to the vents, drawing from both of the worlds simultaneously, can pull off these reactions. The vent animals themselves graze on the bacterial mats, in the case of vent shrimp, or nurture bacteria within themselves, as if tending an internal farm. This explains why, for example, the giant tubeworms don't need a digestive tract; they're fed from inside by herds of bacteria. But the strict requirement to provide both hydrogen sulphide and oxygen gives the animal hosts some interesting dilemmas, for they must bring a little of the two worlds together within themselves. Much of the curious anatomy of the tubeworms stems from this unyielding obligation.

It didn't take long for the conditions in the vent world to register with scientists considering the origin of life, first among them the oceanographer John Baross, at the University of Washington in Seattle. The vents immediately solved many of the problems of the soup, most obviously the problem with thermodynamics; there was nothing of the equilibrium about that black belching smoke. Having said that, the interface between the vents and the oceans would have been rather different on the early earth, as there was little or no oxygen back then. The driving force could not have been the reaction of hydrogen sulphide with oxygen, as in modern respiration. In any case, respiration, at the cellular level, is a complex process that must have taken time to evolve; it can't have been the primordial energy source. Instead, according to the iconoclastic German chemist and patent attorney Günter Wächtershäuser, that earliest engine of life was the reaction of hydrogen sulphide with iron to form the mineral iron pyrites, a reaction that occurs spontaneously, releasing a modicum of energy that can be captured, at least in principle.

Wächtershäuser came up with a chemical scheme for the origin of life that looked like nothing else. The energy released by the formation of pyrites is not enough to convert carbon dioxide into organic matter, so Wächtershäuser hit on carbon monoxide as a more reactive intermediate; this gas is indeed detected in acid vents. He promoted other sluggish organic reactions with various iron–sulphur minerals, which seemed to have outlandish powers of catalysis. And for an encore, Wächtershäuser and his colleagues managed to demonstrate many of these theoretical reactions in the lab, proving them to be more than merely plausible. It was a tour de force that overturned decades-old ideas about how life might have originated, conjuring it up in a hellish environment from the most unexpected of ingredients, essentially hydrogen sulphide, carbon monoxide and iron pyrites – two poisonous gases and fool's gold. One scientist, on first reading Wächtershäuser's work, remarked that it felt like stumbling across a scientific paper that had fallen through a time warp from the end of the twenty-first century.

But is he right? Harsh criticisms have been levelled at Wächtershäuser too, in part because he is a genuine revolutionary, overturning long-cherished ideas; in part, because his haughty manner tends to exasperate fellow scientists; and in part, because there are legitimate misgivings about the picture he paints. Perhaps the most intractable failing is the 'concentration problem', which also afflicts the idea of a soup. Any organic molecules must dissolve in an ocean of water, and so are highly unlikely to ever meet each other and react to form polymers like RNA and DNA. There is nothing to contain them. Wächtershäuser counters that all his reactions can take place on the surface of minerals like iron pyrites. But there is a difficulty with this too, which is that reactions cannot run to completion if the end-products are not released from the surface of the catalyst. Everything either gums up or dissipates.[3]

Mike Russell, now at the Jet Propulsion Laboratory in Pasadena, proposed a solution to all these problems in the mid-1980s. Russell is a kind of prophetic scientific bard, prone to incantations of 'geopoetry', and has a view of life rooted in thermodynamics and geochemistry that seems obscure to many biochemists. But over the decades, Russell's ideas have attracted a growing band of supporters, who see in his vision a uniquely workable solution to the origin of life.

Both Wächtershäuser and Russell agree that hydrothermal vents are central to the origin of life; but beyond that, where one sees black the other sees white; one postulates volcanism, the other its antithesis; one prefers acids, the other alkalis. For two ideas that are sometimes confounded, they have remarkably little in common. Let me explain.

The ocean ridges, which host the black smokers, are the source of new, spreading sea floor. From these centres of volcanic activity, the rising magma slowly forces the adjoining tectonic plates apart, the plates creeping away at the speed of growing toenails. As these inching plates collide with each other, far away, one plate is forced to plunge beneath another, while the other is thrust into petrified convulsions. The Himalayas, the Andes, the Alps, all were thrown into relief by the collision of tectonic plates in this way. But the slow movement of fresh crust across the sea floor also exposes new rocks derived from the mantle, the layer beneath the crust. Such rocks give rise to a second type of hydrothermal vent, very different to the black smokers, and it is this type of vent that Russell himself champions.

This second type of vent is not volcanic, and there's no magma involved. Instead, it depends on the reaction of these freshly exposed rocks with seawater. Water doesn't just percolate into such rocks: it physically reacts with them; it is incorporated into them, altering their structure to form hydroxide minerals like serpentine (named after its resemblance to the mottled green scales of a serpent). The reaction with seawater expands the rock, causing it to crack and fracture, which in turn permits further seawater to penetrate, perpetuating the process. The scale of such reactions is astonishing. The volume of water bound into rock in this way is believed to equal the volume of the oceans themselves. As the ocean floor spreads, these expanded, hydrated rocks ultimately plunge beneath a colliding plate, where they are superheated again in the mantle. Now they give up their water, releasing it into the bowels of the earth. This contamination with seawater drives the convective circulation deep in the mantle, forcing magma back up to the surface at the mid-ocean ridges and volcanoes. And so the turbulent volcanism of our planet is largely driven by a continuous flux of seawater through the mantle. It's what keeps our world out of equilibrium. It is the turning of our globe.[4]

But the reaction of seawater with mantle-derived rocks does more than just drive the relentless volcanism of our planet. It also releases energy as heat, along with copious quantities of gases like hydrogen. In fact, the reaction transfigures everything dissolved in seawater as if it were a magic distorting mirror, reflecting back grotesquely swollen images, in which all the reactants are loaded up with electrons (technically they're said to be 'reduced'). The main gas emanating is hydrogen, simply because seawater is mostly water; but there are smaller amounts of various other gases reminiscent of Stanley Miller's mixture, so useful for generating the precursors of complex molecules like proteins and DNA. Thus carbon dioxide is transformed into methane; nitrogen returns as ammonia; and sulphate belches back as hydrogen sulphide.

The heat and gases make their way back to the surface, where they break through as the second type of hydrothermal vent. These differ in virtually every detail from black smokers. Far from being acidic, they tend to be quite strongly alkaline. Their temperature is warm or hot, but well below the super-heated fury of black smokers. They are usually found some way away from the mid-ocean ridges, the source of fresh spreading sea floor. And rather than forming vertical black chimneys with a single orifice, through which black smoke billows, they tend to form complex structures, riddled with tiny bubbles and compartments, which precipitate as the warm alkaline hydrothermal fluids percolate into the cold ocean waters above. I suspect the reason that only a few people have ever heard of this type of vent relates to the off-putting term 'serpentinisation' (again, from the mineral serpentine). For our purposes, let's simply label them 'alkaline vents', even if this might sound a bit limp compared with the virility of 'black smokers'. We'll see the full significance of the word 'alkaline' later.

Curiously, until recently, alkaline vents were predicted in principle, but only known from a few fossil deposits. The most famous, at Tynagh in Ireland, is about 350 million years old, and set Mike Russell thinking, back in the 1980s. When he examined thin sections of the bubbly rocks from near the fossil vent under an electron microscope, he found the tiny compartments were a similar size to organic cells, a tenth of a millimetre or less in diameter, and interconnected in labyrinthine networks. He postulated that similar mineral cells could be formed when alkaline vent fluids mingled with acidic ocean waters, and soon succeeded in producing porous rocky structures in the lab by mixing alkalis with acids. In a *Nature* letter in 1988, Russell noted that alkaline vent conditions should have made them an ideal hatchery for life. The compartments provided a natural means of concentrating organic molecules, while their walls, composed of iron–sulphur minerals such as mackinawite, endowed these mineral cells with the catalytic properties envisaged by Günter Wächtershäuser. In a 1994 paper, Russell and his colleagues proposed that:

> Life emerged from growing aggregates of iron sulphide bubbles containing alkaline and highly reduced hydrothermal solution. These bubbles were inflated hydrostatically at sulphidic submarine hot springs sited some distance from the oceanic spreading centres 4 billion years ago.

The words were visionary, for at the time such a living deep-sea alkaline vent system had never been discovered. Then, at the turn of the millennium, scientists aboard the submersible *Atlantis* stumbled across exactly this kind of vent, around fifteen kilometres from the mid-Atlantic ridge, on an underwater mountain that also happened to be called the Atlantis massif. Inevitably named Lost City, after the mythic metropolis, its delicate white pillars and fingers of carbonate reaching up into the inky blackness made the name eerily appropriate. The vent field was unlike any other yet discovered. Some of the chimneys were as tall as the black smokers, the largest, dubbed Poseidon, standing sixty metres proud. But far from being stolidly robust structures, these delicate fingers were ornate as gothic architecture, full of 'vacuous doodles' in John Julius Norwich's words. Here, the hydrothermal exhalations were colourless, giving the impression that the city had been suddenly deserted, and preserved for all time in its intricate Gothic splendour. No hellhole of black smokers, these were delicate white non-smokers, their petrified fingers reaching for heaven (see Fig. 1.2).

Invisible they may be, but the exhalations are real enough, and sufficient to support a living city. The chimneys are not composed of iron–sulphur minerals

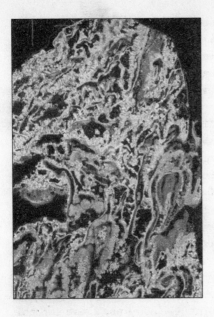

Figure 1.3 Microscopic structure of an alkaline vent, showing interconnecting compartments that provide an ideal hatchery for the origin of life. The section is about one centimetre across and 30 microns thick.

(hardly any iron dissolves in oxygen-rich oceans; Russell's predictions relate to much earlier times), but their structure is still porous, a maze of microscopic compartments with feathery aragonite walls (see Fig. 1.3). Curiously, the old structures that have fallen silent, no longer bubbling with hydrothermal fluids, are far more solid, their pores clogged up with calcite. In contrast, the living vents really are alive, the pores a hive of industrious bacterial activity, in which the chemical disequilibrium is exploited to the full. Animals there are too, rivalling the black smokers in diversity, but falling far short in terms of sheer size. The reason appears to be a matter of ecology. The sulphur bacteria thriving in the black smokers readily adapt to life in their animal hosts, while the bacteria (or strictly archaea) found in the Lost City don't form such partnerships.[5] Lacking internal 'farms' the vent animals grow less efficiently.

Life in the Lost City is built on the reaction of hydrogen with carbon dioxide, which is actually the basis of all life on our planet. In the Lost City, unusually, the reaction is direct, while in practically all other cases, it is indirect. Raw hydrogen, bubbling from the ground as a gas, is a rare gift on our planet, and life is normally obliged to seek out occult supplies, bound in tight molecular grip to other atoms, as in water or hydrogen sulphide. To rip hydrogen from such molecules and bind it on to carbon dioxide costs energy, energy that comes ultimately from the sun in photosynthesis, or from exploiting chemical disequilibria in the vent world. Only in the case of hydrogen gas itself does the reaction take place spontaneously, if painfully slowly. But from a thermodynamic point of view, the reaction is a free lunch that you are paid to eat (in Everett Shock's memorable phrase). In other words, the reaction generates organic molecules directly, and at once releases a substantial amount of energy that can, in principle, be used to power other organic reactions.

So Russell's alkaline vents fit the bill as a hatchery of life. They are an integral part of a system that turns over the surface of the globe, promoting the restless volcanism of our planet. They are perpetually out of equilibrium with the oceans, bubbling a steady supply of hydrogen that reacts with carbon dioxide to form organic molecules. They form a labyrinth of porous compartments, which retain and concentrate any organic molecules formed, making the assembly of polymers, like RNA, far more likely (as we'll see in the next chapter). They are long-lived – the Lost City chimneys have now been venting for 40,000 years, two orders of magnitude longer than most black smokers. And they were more plentiful on the early earth, when the cooling mantle interfaced more directly with the oceans. In those days, too, the oceans were loaded with dissolved iron, and the microcompartments would have had catalytic walls, composed of iron–sulphur minerals, like the fossil vents at Tynagh, in Ireland. They would have worked, in fact, as natural flow reactors, with thermal and electrochemical gradients circulating reactive fluids through catalytic compartments.

That is all very well, but a single reactor, however valuable, scarcely constitutes life. How did life progress from such natural reactors to the complex, marvellous tapestry of invention and ingenuity that we see around us? The answer, of course, is unknown, but there are clues that derive from the properties of life itself, and in particular from an inner core of deeply conserved reactions common to almost all life on earth today. This core of metabolism, a living inner fossil, preserves echoes of the most distant past, echoes that are consonant with primordial origins in an alkaline hydrothermal vent.

There are two ways of approaching the origin of life: the 'bottom up' and the 'top down'. So far in this chapter we have taken the 'bottom up' approach, considering the geochemical conditions and thermodynamic gradients that most likely existed on the early earth. We have come up with warm deep-sea hydrothermal vents, bubbling hydrogen gas into an ocean saturated with carbon dioxide, as the most likely setting for the origin of life. Natural electrochemical reactors would have been capable of generating both organic molecules and energy; but we haven't considered yet exactly what reactions are likely to have taken place, or how they led to life as we know it.

The only true guide to how life came to be is life as we know it today, that is to say, the 'top down' approach. We can catalogue the properties shared by all living things, to reconstruct the hypothetical properties of a Last Universal Common Ancestor, known fondly as LUCA. So, for example, because only a small subset of bacteria is capable of photosynthesis, it is unlikely that LUCA herself was photosynthetic. If she were, then the great majority of her descendants must have abandoned a valuable skill, which seems at best improbable even if it can't be excluded with certainty. Conversely, all life on earth shares common properties: all living things are composed of cells (excluding viruses, which can only operate within cells); all have genes made of DNA; all encode proteins by way of a universal code for particular amino acids. And all living things share a common energy currency, known as ATP (adenosine triphosphate), which functions as a sort of £10 note, capable of 'paying' for all kinds of work about the cell (more on this later). We can infer, reasonably, that all living organisms inherited their shared properties from that remote common ancestor, LUCA.

All life today also shares a common core of metabolic reactions, at the heart of which is a little cycle of reactions known as the Krebs cycle, after Sir Hans Krebs, the German Nobel laureate who first elucidated the cycle in Sheffield in the 1930s, after fleeing the Nazis. The Krebs cycle occupies hallowed ground in biochemistry, but for generations of students it felt like the worst kind of dusty ancient history, to be rote-learned in time for exams and then forgotten.

Yet there is something iconic about the Krebs cycle. Pinned to the walls of cluttered offices in biochemistry departments – the kind of office where you'll find piles of books and papers on the desk, spilling over on to the floor and into the bin, unemptied for a decade – you'll often find a faded, curling, dog-eared metabolic chart. You peer at it with a mixture of fascination and horror while you wait for the professor to return. They are shocking in their complexity, like a lunatic's version of an underground tube map, with little arrows running in all directions, looping back round on each other. Although faded, you can just make out that these arrows are all colour-coded for different pathways, proteins in red, lipids in green, and so on. Down towards the bottom, somehow giving the impression that it is at the centre of this insurrection of arrows, is a tight little circle, maybe the only circle, indeed the only ordered bit, on the whole map. That's it, the Krebs cycle. And as you peer at it you begin to appreciate that virtually all the other arrows on the map somehow spin off from the Krebs cycle, like the spokes of a mangled wheel. It is the centre of everything, the metabolic core of the cell.

The Krebs cycle doesn't feel so dusty any more. Recent medical research has shown it to stand at the heart of the cell's physiology as well as its biochemistry. Changes in the rate at which the cycle spins affect everything from ageing to cancer to energy status. But what came as an even bigger surprise is that the Krebs cycle can go backwards. Normally the cycle consumes organic molecules (from food) and spins off hydrogen (destined for burning with oxygen in respiration) and carbon dioxide. The cycle therefore not only provides precursors for metabolic pathways, it also serves up the little packets of hydrogen needed for generating energy as ATP. In reverse, the Krebs cycle does the opposite: it sucks in carbon dioxide and hydrogen to form new organic molecules, all the basic building blocks of life. And instead of releasing energy as it spins, the reverse cycle consumes ATP. Provide it with ATP, carbon dioxide and hydrogen, and the cycle spins out the basic building blocks of life, as if by magic.

This reverse spinning of the Krebs cycle is not widespread even in bacteria, but it is relatively common in the bacteria that live in hydrothermal vents. It is plainly an important, if primitive, way of converting carbon dioxide into the building blocks of life. The pioneering Yale biochemist, Harold Morowitz, now at the Krasnow Institute for Advanced Study, Fairfax, Virginia, has been teasing out the properties of the reverse Krebs cycle for some years. In broad terms, his conclusion is that, given sufficient concentrations of all the ingredients, the cycle will spin on its own. It is bucket chemistry. If the concentration of one intermediate builds up, it will tend to convert into the next intermediate in succession. Of all possible organic molecules, those of the Krebs cycle are the most stable, and so the most likely to form. In other words, the Krebs cycle was not 'invented' by genes, it is a matter of probabilistic chemistry and thermodynamics. When genes evolved, later on, they conducted a score that already existed, just as the conductor of an orchestra is responsible for the interpretation – the tempo and the subtleties – but not the music itself. The music was there all along, the music of the spheres.

Once the Krebs cycle was spinning and provided with a source of energy, side-reactions would have been almost inevitable, giving rise to more complex precursors, such as amino acids and nucleotides. How much of the core metabolism of life on earth arises spontaneously, and how much is a later product of genes and proteins is an interesting question, and one that is beyond the scope of a book like this. But I would like to make one general point. The great majority of attempts to synthesise the building blocks of life have been too 'purist'. They start with simple molecules like cyanide, which have nothing to do with the chemistry of life as we know it (in fact they are anathema to it), then attempt to synthesise the building blocks of life, by playing around with factors like the pressure, temperature or electrical discharges, totally unbiological parameters, all of them. But what happens when you start out with the molecules of the Krebs cycle and some ATP, ideally in an electrochemical reactor like that proposed by Mike Russell? Just how much of our dog-eared metabolic chart arises spontaneously from these ingredients in a kind of ethereal cast that gradually fills up from the bottom with the thermodynamically most likely molecules? I am not alone in suspecting quite a lot of it, perhaps up to the level of small proteins (strictly, polypeptides) and RNA, at which point natural selection begins to take over.

All this is a matter of experimentation; and these experiments are mostly yet to be done. For any of it to be realistic, we need a nice steady production of that magic ingredient, ATP. And on that score, you may well feel that we're getting ahead of ourselves, trying to run before we've learnt to walk. How do we generate ATP? The answer I find most persuasive comes from the brilliant if frequently outspoken American biochemist, Bill Martin, who forsook the US to take a job as professor of botany at the University of Düsseldorf. From here, Martin has been the source of a steady flow of iconoclastic ideas about the origin of almost everything that matters in biology. Some of it may be wrong, but he is always thrilling and nearly always makes one see biology from a different point of view. A few years ago, Martin sat down with Mike Russell, and the pair of them attacked the transition from geochemistry to biochemistry. Since then the insights have been flowing free. Let's take the ride.

Martin and Russell went back to basics: the flow of carbon into the organic world. Today, they note, there are only five metabolic pathways by which plants and bacteria incorporate hydrogen and carbon dioxide into the living world, to generate organic matter, one of which is the reverse Krebs cycle, as we've seen. Four of the five pathways consume ATP (as does the Krebs cycle), and so can only take place with an input of energy. Yet the fifth pathway, the straight reaction of hydrogen with carbon dioxide, not only produces organic molecules, but it also releases energy. Two groups of ancient organisms do exactly this, via a series of broadly similar steps. And one of these two groups we've already met, the 'archaea' that thrive in the Lost City vent field.

If Martin and Russell are right, the remote ancestors of these archaea were performing the same set of reactions in an almost identical environment, 4,000 million years ago, at the dawn of life. But the reaction of hydrogen with carbon dioxide is not quite as straightforward as it sounds, for the two molecules don't react spontaneously. They are rather 'shy', and need to be persuaded to dance by a catalyst; and they also need a small input of energy to get things going. Only then do the two conjoin, releasing rather more energy as they do so. The catalyst is simple enough. The enzymes that catalyse the reaction today contain little clusters of iron, nickel and sulphur at their cores with a structure very similar to a mineral found in vents. This suggests that the primordial cells simply incorporated a ready-made catalyst, a feature that points to the great antiquity of the pathway, as it does not entail the evolution of sophisticated proteins. As Martin and Russell put it, the pathway has rocky roots.

The source of energy needed to get the ball rolling, in the vent world at least, turns out to be the vents themselves. An unexpected reaction product betrays their hand: a reactive form of vinegar known as an *acetyl thioester*.[6] Acetyl thioesters form because carbon dioxide is quite stable and resists attack by hydrogen, but is vulnerable to more reactive 'free-radical' fragments of carbon or sulphur found in vents. In effect, the energy needed to persuade carbon dioxide to react with hydrogen comes from the vents themselves in the form of reactive free radicals, and these give rise to the acetyl thioesters.

Acetyl thioesters are significant because they stand at an ancient branch point in metabolism, still found in organisms today. When carbon dioxide reacts with an acetyl thioester we take a branch leading to the formation of more complex organic molecules. The reaction occurs spontaneously and releases energy, to produce a three-carbon molecule called pyruvate, a name that ought to make biochemists sit up and blink, for it is an entry point into the Krebs cycle. In other words, a few simple reactions, all thermodynamically favourable, and several catalysed by enzymes with mineral-like clusters at their core, giving them 'rocky roots', take us straight to the metabolic heart of life, the Krebs cycle, without any more ado. And once we've broken into the Krebs cycle, all we need is a nice steady supply of ATP to start it spinning, to generate the building blocks of life.

Energy is exactly what the other prong of the fork provides, when phosphate reacts with another acetyl thioester. Admittedly the reaction doesn't produce ATP, but a simpler form called acetyl phosphate. Even so, this serves much the same purpose and is still used alongside ATP by some bacteria today. It does exactly the same thing as ATP: it transfers its reactive phosphate group on to other molecules, giving them a kind of energy tag that activates them in turn. The process is a bit like the children's game of tag, in which one child is 'it' and must touch a second child, who then becomes 'it' instead. The child who is 'it' gains a 'reactivity' to pass on to the next child. Transferring phosphate from one molecule to another works in much the same way: the reactive tag activates molecules that would not otherwise react. This is how ATP can drive the Krebs cycle backwards, and acetyl phosphate does exactly the same thing. Once the reactive phosphate tag has been transferred, the waste is simply vinegar, a common product of bacteria today. Next time you open a bottle of wine that has gone sour (turning to vinegar) spare a thought for the bacteria at work in the bottle, generating a waste product as old as life itself, a waste more venerable than even the finest vintage.

Pulling all this together, alkaline hydrothermal vents continuously generate acetyl thioesters, providing both a starting point for forming more complex organic molecules and the energy needed to make them, packaged in a format essentially the same as that used by cells today. The mineral cells that riddle the chimneys provide at once the means of concentrating the products, favouring such reactions, and the catalysts needed to speed up the process, without any requirement for complex proteins at this stage. And finally, the bubbling of hydrogen and other gases into the labyrinth of mineral cells means that all the raw materials are replenished continually and thoroughly mixed. It is truly a fountain of life, except for one niggling little detail with the most pervasive consequences.

The problem relates to that little kick of energy that's needed up front to warm relations between hydrogen and carbon dioxide. I mentioned that it's not a problem in the vents themselves, as the hydrothermal conditions form reactive free radicals that get the ball rolling. But it *is* a problem for free-living cells that don't live in vents. Instead they need to spend ATP to get things going, like buying a drink to break the ice on a first date. What's the problem with that? It's a matter of accountancy. The reaction of hydrogen with carbon dioxide releases enough energy to generate one molecule of ATP. But if you must spend one ATP to generate one ATP, then there's no net gain. And if there's no net gain there can't be any spinning of the Krebs cycle, no production of complex organic molecules. Life might be able to get going in the vents, but then should remain forever bound to the vents by a kind of thermodynamic umbilical chord that could never be cut.

Obviously life isn't tied to the vents. If this whole account is not pure make-believe, how did we escape? The answer put forward by Martin and

Russell is marvellous, for it explains why almost all life today makes use of an utterly peculiar method of respiration to generate energy, perhaps the most confoundingly counterintuitive mechanism in all biology.

In *The Hitchhiker's Guide to the Galaxy*, the hopelessly inept ancestors of modern humans crash on to planet earth and supplant the resident apemen. They form a subcommittee to reinvent the wheel and adopt the leaf as legal tender, making everyone enormously rich. But they run into a serious problem with inflation, in which it costs about three deciduous forests to pay for a single ship's peanut. So our ancestors embark on a massive deflation programme and burn down all the forests. It all sounds horribly plausible.

Sinking beneath the frivolity, there is, I suppose, a serious point about the nature of currency – there is nothing whatsoever to anchor value. A peanut could be worth a gold bar, one penny, or three deciduous forests; it all depends on relative valuation, rarity and so on. A £10 note can be worth whatever it wants to be. But this is not the case in chemistry. Earlier on, I compared ATP to a £10 note, and I chose the value with care. The bond energies in ATP are such that you must spend £10 to make one ATP, and you receive exactly £10 when you spend it. It is not relative in the same way as human currency. And that's the root of the problem for any bacteria that try to leave the vents. ATP is not so much a universal currency as a universal £10 note, inflexible in value, and with no such thing as small change. If you want to buy a cheap drink to break the ice on a first date, you must hand over your £10 note and, even if the drink costs £2, you still don't get any change – there is no such thing as a fifth of an ATP molecule. And when you capture the energy released by the reaction of hydrogen with carbon dioxide, you can only store it in units of £10. Let's say you could, in principle, gain £18 from the reaction; that's not sufficient to make two ATPs, so you must make only one. You lose £8 because there is no such thing as small change. Most of us face the same irritating problem at bureaux de change, which only deal in large denominations.

Overall, then, despite needing to spend £2 to get things going, with a payback of £18, when forced to use our universal £10 note, we must spend £10 to gain £10. Bacteria can't avoid this equation: none can grow by the straight reaction of hydrogen with carbon dioxide using ATP alone. And yet they do grow, by way of an ingenious method of breaking up the £10 note into small change, a method known by the formidable name of *chemiosmosis*, which earned its prime expositor, the eccentric British biochemist Peter Mitchell, the Nobel Prize in 1978. The award finally drew to a close decades of bitter disputes. Today, though, with the perspective of another millennium, we can see that Mitchell's discovery ranks among the most significant of the twentieth century.[7] But even those few researchers who long upheld the importance of chemiosmosis struggle to explain why such a strange mechanism should be ubiquitous in life. Like the universal genetic code, the Krebs cycle and ATP, chemiosmosis is universal to all life, and appears to have been a property of the last universal common ancestor, LUCA. Martin and Russell explain why.

In the broadest of terms, chemiosmosis is the movement of protons over a membrane (hence the resemblance in name to osmosis, the movement of water over a membrane). In respiration, what happens is this. Electrons are stripped from food and passed along a chain of carriers to oxygen. The energy released at several points is used to pump protons across a membrane. The outcome is a proton gradient over the membrane. The membrane acts a bit like a hydroelectric dam. Just as water flowing down from a hilltop reservoir drives a turbine to generate electricity, so in cells the flow of protons through protein turbines in the membrane drives the synthesis of ATP. This mechanism was totally unexpected: instead of having a nice straightforward reaction between two molecules, a strange gradient of protons is interpolated in the middle.

Chemists are used to working with whole numbers; it's not possible for one molecule to react with half of another molecule. Perhaps the most confounding aspect of chemiosmosis is that fractions of whole numbers abound. How many electrons need to be transferred to produce one ATP? Somewhere between 8 and 9. How many protons? The most accurate estimate yet is 4.33. Such numbers made no sense at all, until the intermediary of a gradient was appreciated. A gradient, after all, is composed of a million gradations: it doesn't break into whole numbers. And the great advantage of a gradient is that a single reaction can be repeated again and again just to generate one single ATP molecule. If one particular reaction releases a hundredth of the energy needed to generate one ATP, the reaction is simply repeated a hundred times, building up the gradient step by step until the proton reservoir is big enough to generate a single ATP. Suddenly the cell can save up; it has a pocket full of small change.

What does all this mean? Let's go back to the reaction of hydrogen with carbon dioxide. It still costs bacteria one ATP to get the ball rolling; but they are now able to generate more than one ATP, as they can save up towards a second ATP. Not a good living, perhaps, but an honest one. More to the point, it makes the difference between the possibility of growth, and no possibility of growth. If Martin and Russell are right, and the earliest forms of life grew from this reaction, then the only way life could leave the deep-sea vents was by chemiosmosis. It's certainly true that the only forms of life that live from this reaction today both depend on chemiosmosis and can't grow without it. And it's equally true that almost all life on earth shares this same curious mechanism, even though it's not always needed. Why? I imagine simply because they inherited it from a common ancestor that couldn't live without it.

But here is the chief reason to think Martin and Russell are right – the use of protons. Why not, for example, charged sodium, potassium or calcium atoms, which are used by our own nervous systems? There's no obvious reason why protons should be preferred to a gradient of any other type of electrically charged particle; and there are some bacteria that generate a sodium gradient rather than a proton gradient, albeit rarely. The main reason, I think, goes back to the properties of Russell's vents. Recall that the vents bubble alkaline fluids into an ocean that is made acidic by dissolved carbon dioxide. Acids are defined in terms of protons: an acid is rich in protons, an alkali poor. So bubbling alkaline fluids into acidic oceans produces a natural proton gradient. In other words the mineral cells in Russell's alkaline vents are naturally chemiosmotic. Russell himself pointed this out many years ago, but the realisation that bacteria simply could not leave the vents without chemiosmosis was one of the fruits of his collaboration with Martin, who looked into the energetics of microbes. And so these electrochemical reactors not only generate organic molecules and ATP, they even handed over an escape plan, the way to evade the universal £10 note problem.

Of course, a natural proton gradient is only of use if life is able to harness the gradient, and later on generate its own gradient. While it's certainly easier to harness a pre-existing gradient than it is to generate something from scratch, neither is straightforward. These mechanisms evolved by natural selection, there is no doubt. Today it requires numerous proteins specified by genes, and there is no reason to suppose that such a complex system could have evolved in the first place without proteins and genes – genes composed of DNA. And so we have an interesting loop. Life could not leave the vents until it had learnt how to harness its own chemiosmotic gradient, but it could only harness its own gradient using genes and DNA. It seems inescapable: life must have evolved a surprising degree of sophistication in its rocky hatchery.

This paints an extraordinary portrait of the last common ancestor of all life on earth. If Martin and Russell are right – and I think they are – she was not a free-living cell but a rocky labyrinth of mineral cells, lined with catalytic walls composed of iron, sulphur and nickel, and energised by natural proton gradients. The first life was a porous rock that generated complex molecules and energy, right up to the formation of proteins and DNA itself. And that means we have only followed half the story in this chapter. In the next, we'll consider the other half – the invention of that most iconic of all molecules, the stuff of genes, DNA.

DNA

The Code of Life

There's a blue plaque on the Eagle pub in Cambridge, mounted in 2003 to commemorate the fiftieth anniversary of an unusual turn in pub conversation. At lunchtime on 28 February 1953, a couple of regulars, James Watson and Francis Crick, burst into the pub and announced that they had discovered the secret of life. Although the intense American and voluble Brit with an irritating laugh must at times have seemed to verge on a comic double-act, this time they were serious – and half right. If life can be said to have a secret, it is certainly DNA. But Crick and Watson, for all their cleverness, then only knew half its secret.

That morning Crick and Watson had figured out that DNA is a double helix. An inspired leap of mind based on a mixture of genius, model-building, chemical reasoning, and a few pilfered X-ray diffraction photos, their conception was, in Watson's words, just 'so pretty it had to be right'. And the more they talked that lunchtime the more they knew it was. Their solution was published in *Nature* on 25 April as a one-page letter, a kind of announcement not unlike the birth notices in a local newspaper. Unusually modest in tone (Watson famously wrote that he had never met Crick in a modest mood, and he wasn't much better himself), the paper closed with the coy understatement: 'It has not escaped our notice that the specific pairing we have postulated immediately suggests a possible copying mechanism for the genetic material.'

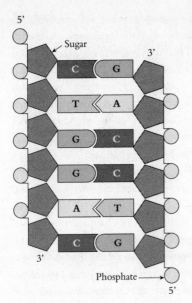

Figure 2.1 Base pairs in DNA. The geometry of the letters means that G only binds to C, and A only binds to T.

DNA, of course, is the stuff of genes, the hereditary material. It codes for human being and amoeba, mushroom and bacterium, everything on this earth bar a few viruses. Its double helix is a scientific icon, the two helices pursuing each other round and round in an endless chase. Watson and Crick showed how each strand complements the other at a molecular level. Prise the strands apart and each acts as a template to reform the other, forging two identical double helices where once there was one. Every time an organism reproduces, it passes a copy of its DNA to its offspring. All it needs to do is pull the two strands apart to produce two identical copies of the original.

While the detailed molecular mechanics could give anybody a headache, the principle itself is beautifully, breathtakingly, simple. The genetic code is a succession of letters (more technically 'bases'). There are only four letters in the DNA alphabet – A, T, G and C. These stand for adenine, thymine, guanine and cytosine, but the chemical names needn't worry us. The point is that, constrained by their shape and bond structure, A can only ever pair with T, and C with G (see Fig. 2.1). Prise the double helix apart, and each strand bristles with unpaired letters. For every exposed A, only a T can bind; for every C, a G; and so on. These base pairs don't just complement each other, they really *want* to bind to each other. There's only one thing to brighten up the dull chemical life of a T and that's close proximity to an A. Put them together and their bonds sing in lovely harmony. This is true chemistry: an authentic 'basic attraction'. So DNA is not merely a passive template; each strand exerts a sort of magnetism for its alter ego. Pull the strands apart and they will spontaneously coalesce together again, or, if they're kept apart, each strand is a template with an urgent tug for its perfect partner.

The succession of letters in DNA seems endlessly long. There are nearly 3 billion letters (base pairs) in the human genome, for example – 3 gigabases, in the lingo. That's to say, a single set of chromosomes in the nucleus of a cell contains a list of 3,000,000,000 individual letters. If typed out, the human genome would fill about 200 volumes, each of them the size of a telephone directory. And the human genome is by no means the largest. Rather surprisingly, that record falls to a measly amoeba, *Amoeba dubia*, with a gargantuan genome of 670 gigabases, some 220 times the size of our own. Most of it seems to be 'junk', coding for nothing at all.

Every time a cell divides, it replicates all of its DNA, a process that takes place in a matter of hours. The human body is a monster of 15 *million million* cells, each one harbouring its own faithful copy of the same DNA (two copies in fact). To form your body from a single egg cell, your DNA helices were prised apart to act as a template 15 million million times (and indeed many more, for cells die and are replaced all the time). Each letter is copied with a precision bordering on the miraculous, recreating the order of the original with an error rate of about one letter in 1,000 million. In comparison, for a scribe to work with a similar precision, he would need to copy out the entire bible 280 times before making a single error. In fact, the scribes' success was a lot lower. There are said to be 24,000 surviving manuscript copies of the New Testament, and no two copies are identical.

Even in DNA, though, errors build up, if only because the genome is so very big. Such errors are called point mutations, in which one letter is substituted for another by mistake. Each time a human cell divides, you'd expect to see about three mutations per set of chromosomes. And the more times that a cell divides, the more such mutations accumulate, ultimately contributing to diseases like cancer. Mutations also cross generations. If a fertilised egg develops as a female embryo, it takes about thirty rounds of cell division to form a new egg cell; and each round adds a few more mutations. Men are even worse: a hundred rounds of cell division are needed to make sperm, with each round linked inexorably to more mutations. Because sperm production goes on throughout life, round after round of cell division, the older the man, the worse it gets. As the geneticist James Crow put it, the greatest mutational health hazard in the population is fertile old men. But even an average child, of youthful parents, has around 200 new mutations compared with their parents (although only a handful of these may be directly harmful).[1]

And so despite the remarkable fidelity with which DNA is copied, change happens. Every generation is different from the last, not only because our genes are stirred by sex, but also because we all carry new mutations. Many of these mutations are the 'point' mutations that we've talked about, a change in a single DNA letter, but some are altogether more drastic. Whole chromosomes are replicated or fail to separate; vast tracts of DNA are deleted; viruses insert new chunks; bits of chromosomes invert themselves, reversing the sequence of letters. The possibilities are endless, though the grossest changes are rarely compatible with survival. When seen at this level, the genome seethes like a snakepit, with its serpentine chromosomes fusing and dividing, eternally restless. Natural selection, by casting away all but the least of these monsters, is actually a force for stability. DNA morphs and twists, selection straightens. Any positive changes are retained, while more serious errors or alterations miscarry, literally. Other mutations, less serious, may be associated with disease later in life.

The shifting sequence of letters in DNA is behind almost everything that we read about in the papers concerning our genes. DNA fingerprinting, for example – used to establish paternity, impeach presidents, or incriminate suspects decades after the event – is based on differences in the sequence of letters between individuals. Because there are so many differences in DNA, we each have our own unique DNA 'fingerprint'. Likewise, our susceptibility to many diseases depends on tiny differences in DNA sequence. On average, humans differ by around one letter every 1,000 or so, giving 6 to 10 million single letter differences in the human genome, known as 'snips' (for 'single nucleotide polymorphisms'). The existence of snips means that we all harbour slightly different versions of most genes. While many snips are almost certainly inconsequential, others are statistically associated with conditions like diabetes or Alzheimer's disease, although exactly how they exert their effects is all too often uncertain.

Despite these differences, it's still possible to talk about a 'human genome'; for all the snips, 999 letters out of every 1,000 are still identical in all of us. There are two reasons for this: time and selection. In the evolutionary scheme of things, not a lot of time has passed since we were all apes; indeed a zoologist would assure me we still are. Assuming that humans split off from our common ancestor with chimps around 6 million years ago, and accumulated mutations at the rate of 200 per generation ever since, we've still only had time to modify about 1 per cent of our genome in the time available. As chimps have been evolving at a similar rate, theoretically we should expect to see a 2 per cent difference. In fact the difference is a little less than that; in terms of DNA sequence, chimps and humans are around 98.6 per cent identical.[2] The reason is that selection applies the brakes, by eliminating most of the detrimental changes. If changes are eliminated by selection, the sequences that do persist are obviously more similar to each other than they would be if change were unconstrained; again, selection is straightening.

Going further back into deep time, these two traits, time and selection, conspire to produce the most marvellous and intricate of tapestries. All life on our planet is related, and the readout of letters in DNA shows exactly how. By comparing DNA sequences, we can compute statistically how closely related we are to anything, from monkeys to marsupials, to reptiles, amphibians, fish, insects, crustaceans, worms, plants, protozoa, bacteria – you name it. All of us are specified by exactly comparable sequences of letters. We even share tracts of sequence in common, the bits constrained by common selection, while other parts have altered beyond recognition. Read out the DNA sequence of a rabbit and you will find the same interminable succession of bases, with some sequences identical to ours, others different, intermingling in and out like a kaleidoscope. The same is true of a thistle: the sequence is identical, or similar, in places, but now larger tracts are different, echoing the

vast tracts of time since we shared a common ancestor, and the utterly different ways of life we lead. But our deep biochemistry is still the same. We are
all built from cells that work in much the same way, and these are still specified
by similar sequences of DNA.

Given these deep biochemical commonalities, we would expect to find
sequences in common with even the most remote forms of life, like bacteria,
and so we do. But in fact there is scope for confusion here, as sequence similarity is not plotted on a scale of 0 to 100 per cent, as one might expect, but from
25 to 100 per cent. This reflects the four letters of DNA. If one letter is substituted for another at random, there is a 25 per cent chance that the same letter
will be substituted. Likewise if a random stretch of DNA is synthesised from
scratch in the lab, there must necessarily be a 25 per cent sequence similarity
to any one of our genes chosen at random – the probability that each letter
will match a letter in human DNA is a quarter. As a result, the idea that we are
'half banana' because we share 50 per cent of our genome sequence with a
banana is misleading, to put it mildly. By the same reasoning, any randomly
generated stretch of DNA would be a quarter human. Unless we know what
the letters actually mean, we're completely in the dark.

And that is why Watson and Crick only grasped half of the secret of life
that morning in 1953. They knew the structure of DNA and understood how
each strand of the double helix could serve as a template for another, so
forming the hereditary code for each organism. What they didn't mention in
their famous paper, because it took a further ten years of ingenious research
to find out, was what the sequence of letters actually codes for. Although
lacking the majestic symbolism of the double helix, which cares not a hoot for
the letters entrained in its spiral, deciphering the code of life was perhaps the
larger achievement, one in which Crick himself figured prominently. Most
importantly, from our point of view in this chapter, the deciphered code, initially the most puzzling disappointment in modern biology, gives intriguing
insights into how DNA evolved in the first place, nearly 4 billion years ago.

DNA seems so modern that it's hard to appreciate how little was known about

Figure 2.2 The DNA double helix, showing how the two helices spiral around each other. Separating the strands allows each to act as a template for a new complementary strand.

the rudiments of molecular biology back in 1953. DNA leapt from the original paper by Watson and Crick, its structure rendered schematically by Crick's artist wife Odile as the twisting ladder copied practically without change for half a century (see Fig. 2.2). And Watson's famous book, *The Double Helix*, written in the 1960s, painted a modern view of science. Indeed so influential was it that life probably began to imitate art. I, for one, read Watson's book at school and dreamt of Nobel prizes and seminal discoveries. In retrospect, my sense of how science was actually done was based almost entirely on Watson's book; and my inevitable disillusionment at university may well have stemmed from the failure of reality to live up to the excitement I had anticipated. I took to rock climbing for thrills instead. It was some years before the intellectual excitement of science seeped back into my bones.

But what I actually learned at university, almost all of it, was unknown to Watson and Crick in 1953. It's a commonplace today that 'genes code for proteins', but in the early 1950s there was little agreement even about this. When Watson first arrived in Cambridge in 1951, he was soon exasperated by the open-minded scepticism of Max Perutz and John Kendrew. For them, it had not yet been proved beyond reasonable doubt that genes were composed of DNA rather than proteins. While the molecular structure of DNA was unknown, its chemical composition was plain enough and barely changed from one species to another. If genes were the basis of inheritance and encoded the myriad differences between individuals and species, how could such a monotonous compound, unchanging in composition from animal to plant to bacterium, begin to explain the richness and variety of life? Proteins, in their infinite variety, seemed far more suited to such a monumental task.

Watson himself was among the few biologists to be convinced by the painstaking experiments of the American biochemist Oswald Avery, published in 1944, showing that genes are made of DNA. Only Watson's enthusiasm and belief galvanised Crick to the task in hand – to solve the structure of DNA. Once solved, the question of the code became acute. Again, the depths of ignorance are surprising to a modern generation. DNA is an endless succession of just four different letters, in a seemingly random order. It was easy, in principle, to see that their order could code in some way for proteins. Proteins, too, are composed of a succession of building blocks called amino acids. Presumably the sequence of letters in DNA encoded the sequence of amino acids in proteins. But if the code was universal, as it already appeared to be, the list of amino acids must also be universal. This was by no means established. It had barely even been considered until Watson and Crick sat down in the Eagle to write out the canonical list of 20, found in all textbooks today, over lunch. Remarkably, given that neither of them were biochemists, they got it right first go.

The challenge was now established and swiftly became a mathematical game, constrained by none of the molecular details rote-learned by later generations. Four different letters in DNA had to code for 20 amino acids. That dismissed a direct transliteration; obviously one letter in DNA couldn't stand for one amino acid. It ruled out a doublet code, too, which could specify no more than 16 amino acids (4×4). A minimum of three letters was required: a triplet code (later proved by Crick and Sydney Brenner), in which groups of three DNA letters encoded a single amino acid. But that seemed seriously wasteful. The four letters can be combined into 64 triplets ($4 \times 4 \times 4$) and so could potentially code for 64 distinct amino acids. So why only 20? The magic answer had to make sense of a four-letter alphabet, organised into 64 words of three letters each, coding for 20 amino acids.

Fittingly, perhaps, the first to offer any sort of an answer was not a biologist but the ebullient Russian-born American physicist George Gamow, better known for his theories of the Big Bang. For Gamow, DNA was literally a template for proteins: amino acids nestled into diamond-shaped grooves between the turns of the helix. But Gamow's theory was basically numerological, and he was little perturbed to learn that proteins are not built in the nucleus at all, and so never come into direct contact with DNA. His idea simply became more abstract. In essence, he proposed an overlapping code, which had the great advantage, beloved of cryptographers, of maximising the information density. Imagine a sequence that reads ATCGTC. The first 'word', or, more technically, 'codon', would be ATC, the second TCG, the third CGT, and so on. Critically, overlapping codes always constrain the sequence of amino acids permitted. If ATC codes for a particular amino acid, it must be followed by an amino acid whose codon begins with TC; and the next must begin with a C. When all the permutations are laboriously worked through, a large number of triplets are simply disallowed: they can't be part of this overlapping code because an A must always sit next to a T, a T with a C, and so on. How many triplets are left over to code for amino acids? Exactly 20! said Gamow, with the air of a magician pulling a rabbit from a hat.

It was the first of many clever ideas to be shot down by merciless data. All overlapping codes are sabotaged by their own constraints. For a start, they stipulate that certain amino acids must always appear next to each other in proteins, but Fred Sanger, the quiet genius who went on to win two Nobel prizes, one for sequencing proteins and the other for sequencing DNA, was at the time busy sequencing insulin. It soon became clear that any amino acid can appear next to any other: protein sequence is not constrained at all. A second big problem was that any point mutation (in which one letter is swapped for another) is obliged to affect more than one amino acid in an overlapping code, but experimental data showed that often only a single amino acid is altered. Plainly the real code did not overlap. Gamow's overlapping codes were disproved long before the real code was known. Cryptographers were already beginning to suspect that Mother Nature had missed out on a trick or two.

Crick himself was next up, proposing an idea so beautiful that it was immediately embraced by everyone, despite his own concerns about the lack of supporting data. Crick made use of new insights emerging from several molecular biology labs, notable among them Watson's own new lab at Harvard. Watson had become obsessed with RNA, a shorter single-stranded version of DNA, found in the cytoplasm as well as the nucleus. Better still, thought Watson, RNA was an integral part of the tiny machines, now known as ribosomes, that seemed to be the site of protein synthesis. So DNA sits in the nucleus, inert and immobile. When a protein is needed, a section of DNA is used as a template to make an RNA copy, which then moves physically out of the nucleus to the ribosomes waiting outside. This winged messenger was soon dubbed messenger RNA, or mRNA. So, as Watson put it in a letter to Crick as early as 1952, 'DNA makes RNA makes protein'. The question that interested Crick was this: how is the exact sequence of letters in messenger RNA translated into the sequence of amino acids in a protein?

Crick thought about it, and suggested that an RNA message could be translated with the aid of a series of distinct 'adaptor' molecules, one for each amino acid. These, too, would need to be made of RNA, and each one would have an 'anti-codon', which could recognise and bind to a codon in messenger RNA. The principle, said Crick, is exactly the same as DNA: C pairs with G, A with T, and so on.[3] The existence of such molecular adaptors was purely hypothetical at that juncture, but they were duly discovered within a few years and were composed of RNA, as Crick had predicted. They're now called transfer RNAs, or tRNAs. The whole edifice was beginning to feel like Lego, with pieces joining and detaching, to form marvellous, if fleeting, structures.

Here is where Crick went wrong. I'm describing it in some detail because, although the reality is somewhat more outlandish than Crick anticipated, his ideas may still have some bearing on how it all began in the first place. Crick pictured messenger RNA just sitting in the cytoplasm with its codons projecting like a sow's nipples, each one ready to bind its transfer RNA like a a suckling pig. Eventually, all the tRNAs would nestle up, side by side down the full length of the messenger RNA, with their amino acids projecting out like the tails of the piglets, ready to be zipped up together in a protein.

The problem, for Crick, was that the tRNAs would arrive at random, as they appeared on the scene, and attach themselves to the nearest available codon. But if they didn't start at the start and finish at the finish, how would they know where one codon started and the next ended? How could they locate the correct reading frame? If the sequence reads ATCGTC, as before, one tRNA could bind to ATC and the next to GTC; but what was to stop a tRNA recognising the CGT in the middle of the sequence and binding to that instead, so confounding the whole message? Crick's authoritarian answer was to disallow it. If the message as a whole is to be read without ambiguity, not all codons could make sense. Which ones had to be disallowed? Plainly sequences composed only of A, C, U or G were disqualified; in a string of AAAAAA there would be no way to locate the correct reading frame. Crick then cycled through all the possible combinations of letters. In short, if ATC made sense, then all cyclic permutations of these three letters had to be disallowed (so if ATC is allowed, TCA and CAT must be barred). How many possibilities did that leave? Precisely 20 again! (Of the 64 possible codons, AAA, UUU, CCC and GGG are ruled out, leaving 60; then if only one cyclic permutation in 3 is permitted, 60 divided by 3 is 20.)

Unlike the overlapping codes, Crick's code put no constraints on the order of amino acids in a protein; nor does a point mutation necessarily alter two or three amino acids. When proposed, then, it resolved the frame-reading problem beautifully, reducing 64 codons to 20 amino acids in a numerologically satisfying way; and was fully consistent with all known data. But still it was wrong. Within a few years, it turned out that a synthetic RNA composed purely of AAAs (banned by Crick) did after all code for the amino acid lysine and could be converted into a protein polymer composed entirely of lysine.

As experimental tools grew more sophisticated, several research groups gradually pieced the real code together in the mid-1960s. After the sustained cryptographic effort to crack the code, the apparently haphazard reality came as the most perplexing anticlimax. Far from an elegant numerological resolution, the code is simply degenerate (which is to say, full of redundancy). Three amino acids are encoded by as many as six different codons, others by only one or two. All the codons have a use, three stipulating 'stop here', all the rest of them encoding an amino acid. There seemed to be no order, no beauty; indeed it is the perfect antidote to the idea that beauty is any guide to truth in science.[4] There didn't even seem to be any particular structural reason that could account for the code; there was no strong chemical or physical affiliations between amino acids and particular codons.

Crick declared the disappointing code to be a 'frozen accident', and most could only nod and agree. It was frozen, he said, because any infringements – unfreezing the code – would have serious consequences. A single point mutation changes an amino acid here or there, whereas any alteration in the code itself would induce a catastrophic change in absolutely everything. The distinction is between the occasional typo in a book, which doesn't alter the meaning much, and transforming the entire alphabet into gobbledegook. Once set in stone, then, said Crick, any further tampering with the code would be punishable by death, a view that still resonates with biologists widely today.

But the 'accidental' nature of the code gave Crick a problem. Why just one accident, why not several? If the code is arbitrary, no one code would have any particular benefit over any other. There would then be no reason for a selective 'bottleneck', in which one version of the code had, in Crick's words, 'such a selective advantage over all its competitors that it alone survived'. But if there were no bottleneck, Crick wondered, why didn't several codes coexist in different organisms?

The obvious answer is that all organisms on earth descend from a common ancestor, in which the code was already fixed. More philosophically, life arose just once on earth, making it look unique and improbable, perhaps even a freak event. For Crick, that suggested an infection – a single inoculation. Life, he speculated, was 'seeded' on earth as a bacterial clone derived from a single extraterrestrial organism. He went further, arguing that bacteria had been deliberately seeded by some alien intelligence from a spaceship sent to earth, an idea he called 'directed panspermia'. He developed the theme in a book, *Life Itself*, published in 1981. As Matt Ridley put it, in his superb biography of Crick: 'The subject matter raised not a few eyebrows. The great Crick writing about alien life forms seeding the universe from spacecraft? Had success gone to his head?'

Whether the notion of an accidental code really justifies such weighty philosophy is a matter of opinion. The code itself didn't need to offer any particular advantages or disadvantages to be thrust through a bottleneck: strong selection for any trait at all, even a freak accident such as an asteroid impact, could have wiped out all but a single clone, which by definition would have had but a single code. Be that as it may, Crick's timing was unfortunate. Since the early 1980s, when Crick was writing, we have come to realise that the code of life is neither frozen nor accidental. There are covert patterns in the code, a 'code within the codons', giving a clue to its origins nearly 4 billion years ago. And we now know that the code is a long way from the contemptible cipher condemned by cryptographers, but a code in a million, able to resist change and speed up the pace of evolution at once.

A code within the codons! Various patterns had been discerned in the code from the 1960s on, but most of them were easy to dismiss as little more than a curiosity, mere statistical noise, as indeed did Crick himself. Even when taken together, the overall pattern seemed to make little sense. Why it made so little sense is a good question, addressed by the Californian biochemist Brian K. Davis, who has a long-standing interest in the roots of the genetic code. Davis notes that the very idea of a 'frozen accident' dispelled interest in the origins of the code: why study an accident? Accidents just happen. Beyond that, Davis believes, the handful of researchers who did retain an interest were misled by the prevailing idea of a primordial soup. If the code originated in a soup, then it should have its deepest roots in the molecules most likely to be generated by the physical and chemical processes going on in the soup. And that predicted a core of amino acids as the basis of the code, with others being added on later. There was just enough truth in this idea for the evidence to be tantalising, if confounding. It's only when we see the code as the product of *biosynthesis* – the product of cells capable of making their own building blocks from hydrogen and carbon dioxide – that the patterns begin to make sense.

So what are these elusive patterns? A different pattern is linked with each letter of the triplet code. The first letter is the most striking, for it is associated with the steps that turn a simple precursor into an amino acid. The principle is so startling that it's worth spelling out briefly. In cells today, amino acids are made via a series of biochemical steps, starting from several different simple precursors. The surprising observation is that there is a tie between the first letter of the triplet codon and these simple precursors. Thus, all amino acids formed from the precursor *pyruvate* share the same first letter in the codon – T in this case.[5] I use the example of pyruvate because it is a molecule that we have already met in Chapter 1. It can be formed in hydrothermal vents from carbon dioxide and hydrogen, catalysed by minerals found in the vents. But pyruvate is not alone in this regard. All the amino acid precursors are part of the core biochemistry of all cells, the Krebs cycle, and should form in the type of hydrothermal vent discussed in Chapter 1. The implication, admittedly weak at this point but set to deepen, is that there is a link between hydrothermal vents and the first position of the triplet code.

What about the second letter? Here the association is with the degree to which an amino acid is soluble or insoluble in water, which is to say, its hydrophobicity. Hydrophilic amino acids dissolve in water, whereas hydrophobic amino acids are immiscible, dissolving in fats or oils such as the lipid membranes of cells. The amino acids can be sorted into a spectrum, running from 'very hydrophobic' to 'very hydrophilic', and it's this spectrum that bears a relationship with the second position of the triplet code. Five of the six most hydrophobic amino acids have T as the middle base, whereas all the most hydrophilic have A. The intermediates have a G or a C. Overall, then, there are strong deterministic relationships between the first two positions of every codon and the amino acid encoded, for whatever reasons.

The final letter is where the degeneracy lies, with eight amino acids having (a lovely technical term, this) *fourfold degeneracy*. While most people might picture a fourfold degenerate as a staggering drunkard, who manages to collapse into four different gutters, biochemists merely mean that the third position of the codon is information-free: it doesn't matter which base is present, as all four possibilities encode the same amino acid. In the case of glycine, encoded by the triplet GGG, for example, the final G can be switched to a T, A or C – each triplet still codes for glycine.

The degeneracy of the code in the third position has several interesting implications. We've already noted that a doublet code could encode up to 16 out of the 20 different amino acids. If we eliminate the 5 most complex amino acids (leaving 15, plus a stop codon), the patterns in the first two letters of the code become even stronger. It might be, then, that the primordial code was a doublet, and was only later expanded into a triplet code, by 'codon capture'; the amino acids competed among themselves for the third position. If so, the earliest amino acids may have had an 'unfair' advantage in 'taking over' triplet codons, and this seems to be true. For example, the 15 amino acids most likely to have been encoded by the early doublet code hog between them 53 out of 64 possible triplets, an average of 3.5 codons per amino acid. In contrast, the 5 'later' additions muster only 8 codons between them, an average of just 1.6. It certainly looks like the early birds got the worms.

So let's entertain the possibility that the code was initially a doublet, not a triplet, encoding a total of 15 amino acids (plus one 'stop' codon). This early code seems to have been almost entirely deterministic, which is to say that it was dictated by physical and chemical factors. There are few exceptions to the rules that the first letter is allied to the precursor, while the second letter is linked to the hydrophobicity of the amino acid. There is little scope here for any play of chance, no freedom from the physical rules.

But the third letter is a different matter. Here, with so much flexibility, there could have been a play of chance, and with that it became possible for selection to 'optimise' the code. This, at any rate, was the radical proposition of two English molecular biologists, Lawrence Hurst and Stephen Freeland, in the late 1990s. The pair made science headlines when they compared the genetic code with millions of random computer-generated codes. They considered the damage that could be done by point mutations, in which one letter of a codon is switched for another. Which code, they wondered, could resist such point mutations best, either by retaining exactly the same amino acid, or by substituting a similar one? They found that the real genetic code is startlingly resistant to change: point mutations often preserve the amino acid sequence, and if a change does occur, a physically related amino acid tends to be substituted. In fact, Hurst and Freeland declared the genetic code to be better than a million alternative randomly generated codes. Far from being the folly of nature's blind cryptographer, the code is one in a million. Not only does it resist change, they say, but also by restricting the catastrophic consequences of the changes that do occur, the code actually speeds up evolution: obviously, mutations are more likely to be beneficial if they are not catastrophic.

Short of positing celestial design, the only way to explain optimisation is via the workings of selection. If so, the code of life must have evolved. Certainly, a number of trivial variations in the 'universal' code among bacteria and mitochondria do show that, if nothing else, the code can evolve, at least under exceptional circumstances. But how does it change without causing mayhem, you may ask, with Crick? The answer is: discretely. If an amino acid is encoded by four or even six different codons, some tend to be used more often than others. The rarely used codons can in practice be redesignated to a different (but probably related) amino acid without catastrophic consequences. And so the code evolves.

Altogether, then, the 'code within the codons' speaks of a physical process, initially related to the biosynthesis and solubility of amino acids, followed later by expansion and optimisation. The question is, what kind of physical process did selection begin to act upon?

The answer is not known with any certainty, and there are some stumbling blocks. One of the earliest was the chicken-and-egg problem of DNA and proteins. The problem was that DNA is more or less inert, and requires specific proteins even to replicate itself. On the other hand, specific proteins don't get to be specific by chance. They evolve by natural selection, and for that to happen their structure must be both inheritable and variable. Proteins don't act as their own heritable template: they are coded by DNA. And so proteins can't evolve without DNA, while DNA can't evolve without proteins. If neither can evolve without the other, selection can never get started.

Then in the mid-1980s came the startling discovery that RNA acts as a catalyst. RNA rarely forms a double helix, instead forming smaller molecules with complex shapes that lend themselves to catalysis. And so RNA breaks the loop. In a hypothetical 'RNA world', it takes over the role of both proteins and DNA, catalysing its own synthesis, along with many other reactions. Suddenly there was no need for the code to be all about DNA: it could have grown from the direct interactions of RNA with proteins.

That made good sense in terms of how modern cells work. In cells today, there are no direct interactions between DNA and amino acids; but during protein synthesis, many fundamental reactions are catalysed by RNA enzymes, known as ribozymes. The term 'RNA world' was coined by Watson's Harvard colleague Walter Gilbert, in one of the most widely read *Nature* articles ever written. The idea had a mesmeric effect on the field, while the quest for the code of life was recast from 'how does DNA code for proteins' to 'what kinds of interactions must have taken place between RNA and amino acids'. Yet still the answer was far from obvious.

Given the keen interest in an RNA world, it's perhaps surprising that the catalytic properties of smaller RNA fragments have been largely ignored. If large RNA molecules can catalyse reactions, it's likely that smaller fragments – individual 'letters' or pairs of 'letters' – could also catalyse reactions, albeit less vigorously. Recent research from the redoubtable American biochemist Harold Morowitz, working with molecular biologist Shelley Copley and physicist Eric Smith, suggests exactly this possibility. Their ideas might not be right, but I think this is the kind of theory we should be looking for to explain the origin of life's code.

Morowitz and colleagues postulate that pairs of letters (technically 'dinucleotides') did act as catalysts. They pictured a dinucleotide binding to an amino acid precursor, like pyruvate, and catalysing its conversion into an amino acid. Exactly which amino acid is formed depends on the letters paired together in the dinucleotide (following the code within the codons discussed earlier). In essence the first position specifies the amino acid precursor, the second the type of transformation. If the two letters are UU for example, pyruvate binds and is converted into the fairly hydrophobic amino acid leucine. Morowitz supported this pleasingly simple idea with some clever reaction mechanisms, making it sound at least plausible, although I should like to see some evidence in a test tube that the proposed reactions really do take place.

Just two more steps are needed to arrive at a triplet code from here, at least in principle, and neither assumes anything more than normal pairing between letters. In the first step, an RNA molecule binds to the two-letter dinucleotide via standard base-pairing: G with C, A with U, and so on. The amino acid transfers over to the larger RNA, which being larger has stronger powers of attraction.[6] The outcome is an RNA bound to an amino acid, the identity of which depends on the letters that had formed the dinucleotide. It is, in effect, a prototype of Crick's 'adaptors', an RNA charged with the 'correct' amino acid.

The final step converts a two-letter code into a three-letter code, and again could depend on no more than standard base pairing between RNAs. If such interactions function better with three letters than two (perhaps because the spacing is better, or the binding strength), then we easily switch to a triplet code, in which the first two letters are specified by the constraints of synthesis, while the third letter can vary within limits, allowing optimisation of the code at a later stage. This is where I suspect that Crick's original conception of RNAs nestling up like piglets on a sow's nipples may have been right: spatial constraints could push adjacent RNAs to an 'average' of three letters apart. Notice that as yet there is no reading-frame, no proteins, merely amino acids interacting with RNAs. But the basis of the code is already in place, and additional amino acids can be added in at a later stage by capturing vacant triplet codons.

This whole scenario is speculative, to be sure, and as yet there is little evidence to back it up. Its great virtue is that it sheds light on the origin of the code, taking us from simple chemical affiliations to a triplet codon in a plausible and testable way. Even so, you may feel that this is all very well, but that I'm brandishing RNA around as if it grows on trees. For that matter, how do we switch from simple chemical affiliations to selection for proteins? And how do we get from RNA to DNA? As it happens, there are some striking answers, backed up by surprising findings in the last few years. Gratifyingly, the new findings square beautifully with the idea of life evolving in hydrothermal vents, the setting of Chapter 1.

The first question is, where does all the RNA come from? Despite two decades of intense research into the RNA world, this question was rarely asked in a serious way. The unspoken and frankly nonsensical assumption is that it was somehow just 'sitting there' in a primordial soup.

I'm not being snide here. There are many specific problems in science, and they can't all be answered at once. The wonderful explanatory power of the RNA world depends on a 'given': the prior existence of RNA. For the pioneers of the RNA world, it was not important where RNA came from; the question driving research was, what could it do? Certainly, others were interested in the synthesis of RNA; but they were split in interminably argumentative factions, bickering among themselves about their favoured hypotheses. Perhaps RNA was synthesised in outer space from cyanide; perhaps it was made here on earth by lightning striking methane and ammonia; perhaps it was forged on fool's gold in a volcano. All these scenarios offered a few advantages, but all suffered from the same basic problem, the 'concentration problem'.

It's hard enough to make individual RNA letters (nucleotides) but they will only join together in a polymer (a proper RNA molecule) if the nucleotides are present at high concentration. If present in bulk, nucleotides condense spontaneously into long chains. But if the concentration is low, the opposite happens: RNA breaks back down into its constituent nucleotides. The trouble is that every time an RNA replicates itself, it consumes nucleotides, so lowering their concentration. Unless the pool of nucleotides is replenished continuously, and faster than it is consumed, the RNA world could never work, for all its explanatory power. That would never do. So, for those who just wanted to get on with some productive science, it was best to take RNA as a given.

They were right to do so for the answer was a long time in the coming, if ultimately emerging in dramatic fashion. It's true that RNA doesn't grow on trees, but it does grow in vents, or at least in simulated vents. In an important theoretical paper of 2007, indefatigable geochemist Mike Russell (whom we met in Chapter 1), working with Dieter Braun and his colleagues in Germany, reported that nucleotides should accumulate to extreme levels in vents. The reason relates to the strong thermal gradients that develop there. Recall from Chapter 1 that alkaline hydrothermal vents are riddled with interconnecting pores. Thermal gradients produce two types of current, which circulate through these pores, convection currents (as in a boiling kettle) and thermal diffusion (the dissipation of heat into cooler waters). Between them, these two thermal currents gradually silt up the lower pores with many small molecules, including nucleotides. In their simulated hydrothermal system, the concentration of nucleotides reached thousands and even millions of times the starting level. Such high levels should comfortably condense nucleotides into RNA or DNA chains. As the authors concluded, these conditions provide 'a compelling high-concentration starting point for the molecular evolution of life'.

But that isn't all the vents can do. Longer RNA or DNA molecules theoretically accumulate to even higher levels than single nucleotides: their greater size makes them more likely to silt up in the pores. DNA molecules composed of 100 base pairs are predicted to accumulate at fantastic levels, up to a million billion times the starting concentration. Such high concentrations should in principle enable all the types of interaction we've been discussing, such as the binding of RNA molecules to each other, and so on. Even better, oscillating temperatures (thermal cycling) promotes RNA replication in the same way as the ubiquitous lab technique PCR (the polymerase chain reaction). In PCR, high temperatures unravel DNA, enabling it to act as a template, while condensing in cooler temperatures permits the complementary strand to polymerise. The outcome is an exponential rate of replication.[7]

Taken together, thermal gradients should concentrate single nucleotides to extreme levels in vents, promoting the formation of RNA. Then these same gradients should concentrate RNA, fostering physical interactions between molecules. Finally the oscillating temperatures should promote RNA replication. It's hard to imagine a better setting for the primordial RNA world.

So what about our second question: how do we go from replicating RNAs, competing among themselves, to a more sophisticated system in which RNA begins to code for proteins? Again, vents may hold the answer.

Put RNA in a test tube, along with the raw materials and energy (as ATP) it needs, and it will replicate. In fact, it won't just replicate, but as molecular biologist Sol Spiegelman and others discovered in the 1960s, it will *evolve*. Over test-tube generations, RNA gets faster and faster at replicating, finally becoming monstrously efficient. It becomes Spiegelman's monster – a prolifically replicating strand of RNA, capable only of the most artificial and frenzied existence. Curiously, it doesn't matter where the starting point is: you can start out with a whole virus or with an artificial length of RNA. You can even begin with a mixture of nucleotides and a polymerase to zip them up together. Wherever you start, there is always a tendency to home in on the same 'monster', the same frenetically replicating strand of RNA, barely fifty letters long, Spiegelman's monster. It's a molecular groundhog day.

The point is that Spiegelman's monster does not become more complex. The reason it ends up as a stretch of fifty letters is that this is the binding sequence for the replicase enzyme, without which the strand could not replicate at all. Effectively, RNA can't see past its own nose and is never going to generate complexity in a solution. So how and why did RNA begin to code for proteins, at the cost of its own replicative speed? The only way out of this loop is for selection to occur at a 'higher level', for RNA to become part of a larger entity, which is now the unit of selection, a cell for example. The trouble is that all organic cells are far too complex to just pop into existence without evolution, which is to say, there must be selection for the traits that make a cell, rather than selection for the speed of RNA replication. This is a chicken-and-egg situation just as ineluctable as the DNA–protein loop, albeit less celebrated.

We've seen that RNA breaks the DNA–protein loop beautifully; but what breaks the selection loop? The answer is staring us in the face: it's the ready-made inorganic cells in hydrothermal vents. Such cells are about the same size as organic cells and are formed all the time in active vents. So, if the contents of a cell are especially good at regenerating the raw materials needed to replicate themselves, the cell begins to replicate itself, budding off into new inorganic cells. In contrast, 'selfish' RNAs, which replicate themselves as fast as possible, start to lose out, as they are unable to regenerate the raw materials needed to sustain their own replication.

In other words, in the vent environment, selection gradually shifts from the replicative speed of individual RNA molecules, to the overall 'metabolism' of cells, acting as individual units. And proteins, above all else, are the masters of metabolism. It was inevitable that they would ultimately supplant RNA. But of course proteins didn't come into existence at once; it's likely that minerals, nucleotides, RNAs, amino acids and molecular complexes (amino acids binding to RNA, for example) all contributed to a prototype metabolism. The point is that what began as simple affiliations between molecules became, in this world of naturally proliferating cells, selection for the ability to reproduce the contents of whole cells. It became selection for self-sufficiency, and ultimately for autonomous existence. And ironically, it's in the autonomous existence of cells today that we find our final clue to the origin of DNA itself.

There is among bacteria a deep split. We'll come to appreciate the great significance of this split to our own evolution in Chapter 4. For now, we'll just consider the implications for the origin of DNA, profound enough in themselves. The split is between the eubacteria (from the Greek meaning 'true' bacteria), and a second group, which to all intents and purposes looks just the same. This second group is known as the archaeabacteria, or simply archaea. Their name derives from the idea that they are especially archaic, or ancient, although few people today believe they are any older than the true bacteria.

In fact, in a fluke of fortune bordering on the unbelievable, it might be that both the bacteria and archaea emerged from the very same hydrothermal mound. Little else could explain the fact that they share the same genetic code, as well as many details of protein synthesis, but apparently only learnt to replicate their DNA later on, totally independently. For while DNA and the genetic code certainly evolved only once, DNA replication – the physical mechanism of inheritance in all living cells – apparently evolved twice.

If such a claim came from a lesser man than Eugene Koonin, a meticulous and intellectual Russian-born American computational geneticist at the National Institutes of Health in the US, I would beg leave to doubt it. But Koonin and his colleagues did not set out to prove a radical conception. They stumbled across it in the course of a systematic survey of DNA replication in bacteria and archaea. From detailed gene-sequence comparisons Koonin and his colleagues found that bacteria and archaea broadly share the same mechanisms of protein synthesis. For example, the way in which DNA is read off into RNA, and then RNA translated into proteins, is basically similar, with bacteria and archaea using enzymes that (from their gene sequences) are obviously inherited from a common ancestor. Yet this is far from the case for the enzymes needed for DNA replication. Most of them have nothing at all in common. This curious state of affairs could be explained just by their deep divergence, but then the question arises, why did the equally deep divergence of DNA transcription and translation *not* lead to such an utter dissimilarity? The simplest explanation is Koonin's own radical supposition: DNA replication evolved twice, once in the archaea and once in the bacteria.[8]

Such a claim must have seemed outrageous to many, but to a brilliant and amiably 'ornery' Texan working in Germany it was just what the doctor had ordered. Biochemist Bill Martin, whom we met in Chapter 1, had already teamed up with Mike Russell to explore the origins of biochemistry in hydrothermal vents. Flying in the face of conventional wisdom, in 2003 they wrote up their own proposition: that the common ancestor of bacteria and archaea was not a free-living organism at all, but a replicator of sorts, confined to porous rock; it had not yet escaped the mineral cells riddling hydrothermal mounds. To support their case, Martin and Russell produced a list of other abysmal differences between bacteria and archaea. In particular, their cell membranes and walls are wholly different, implying that the two groups emerged independently from the same rocky confines. The proposal was too radical for many, but for Koonin it fit the observations like a glove.

It didn't take long for Martin and Koonin to put their heads together to consider the origin of genes and genomes in hydrothermal vents, and their stimulating thoughts on the subject were published in 2005. They suggested that the 'life cycle' of mineral cells might have resembled modern retroviruses, such as HIV. Retroviruses have a tiny genome, encoded in RNA rather than DNA. When they invade a cell, retroviruses copy their RNA into DNA using an enzyme called 'reverse transcriptase'. The new DNA is first incorporated into the host's genome, and then read off along with the host cell's own genes. So when forming multiple copies of itself, the virus works from DNA; but while packaging itself up for the next generation, it relies on RNA to transmit the hereditary information. What it lacks, notably, is the ability to replicate DNA, which is in general rather a cumbersome procedure, requiring a number of enzymes.

There are both advantages and disadvantages to such a life cycle. The big advantage is speed. By taking over a host cell's machinery for transcribing DNA into RNA, and translating RNA into proteins, retroviruses rid themselves of the need for a large number of genes and so save themselves a good deal of time and trouble. The big disadvantage is that they depend entirely on 'proper' cells for their existence. A second, less obvious, disadvantage is that RNA is poor at storing information in comparison with DNA. It is chemically less stable, which is to say it is more reactive than DNA. That, after all, is how RNA catalyses biochemical reactions. But this reactivity means that large RNA genomes are unstable and break down, which imposes a maximum size limit well below that needed for independent existence. A retrovirus is, in fact, already nearly as complex as an RNA-encoded entity can be.

But not in mineral cells. The mineral cells offer two advantages, which enable more complex RNA life forms to evolve. The first is that many of the properties required for an independent existence are provided for free in vents, giving the cells a head start: the proliferating mineral cells already provide bounding membranes, energy, and so on. There is a sense, then, in which the self-replicating RNAs that populate the vents are already 'viral'. The second advantage is that 'swarms' of RNAs are constantly mixing and matching through the interconnecting cells; and groups that 'cooperate' well can be selected together if they diffuse in concert to populate newly forming cells.

And so Martin and Koonin envisaged populations of cooperative RNAs emerging in mineral cells, each RNA encoding a handful of related genes. The drawback to this arrangement, of course, is that the RNA populations would be vulnerable to remixing into different, possibly ill-suited, combinations. A cell that managed to hold its 'genome' together, by converting a group of cooperative RNAs into a single DNA molecule, would retain all its advantages. Its replication would then be similar to a retrovirus, its DNA transcribed into a swarm of RNAs that infect adjacent cells, bestowing on them the same ability to deposit information back into a DNA bank. Each new flurry of RNAs would be freshly minted from the bank, and so less likely to be riddled with errors.

How hard would it have been for mineral cells to 'invent' DNA in these circumstances? Not so hard, probably; much easier, in fact, than inventing a whole system for replicating DNA (rather than RNA). There are just two tiny chemical differences between RNA and DNA, but together they make an immense structural difference: the difference between coiled catalytic molecules of RNA, and the iconic double helix of DNA (as predicted, incidentally, by Crick and Watson in their original 1953 *Nature* paper).[9] Both of these tiny changes would be hard to stop taking place virtually spontaneously in vents. The first is the removal of a single oxygen atom from RNA (ribonucleic acid) to give *deoxy*-ribonucleic acid, or DNA. The mechanism today still involves the kind of reactive (technically free-radical) intermediates found in vents. The second difference is the addition of a 'methyl' (CH_3) group on to the letter uracil, to give thymine. Again, methyl groups are reactive free-radical splinters of methane gas, plentiful in alkaline vents.

So making DNA could have been relatively easy: it would have formed as 'spontaneously' in the vents as RNA (I mean its formation from simple precursors would have been catalysed by minerals, nucleotides, amino acids, and so on). A slightly more difficult trick would have been to retain the coded message, which is to say, to make an exact copy of the sequence of letters in RNA in the form of DNA. Yet here too the void is not insuperable. To convert RNA to DNA requires just one enzyme: a reverse transcriptase, held in trust by retroviruses like HIV today. How ironic that the one enzyme that 'breaks' the central dogma of molecular biology – DNA makes RNA makes protein – should have been the enzyme that turned a porous rock infested with viral RNA into life as we know it today! It may be that we owe the very birth of cells to the humble retrovirus.

There is much in this tale left unsaid, many puzzles skipped over in an attempt to reconstruct a story that makes some kind of sense, at least to me. I can't pretend that all of the evidence we've discussed here is conclusive, or much more than clues to the deepest past. Yet they are genuine clues, which will need to be explained by whatever theory turns out to be true. There really are patterns in the code of life, patterns that imply the operation of both chemistry and selection. Thermal currents in deep-ocean vents really do concentrate nucleotides, RNA and DNA, turning their riddling mineral cells into an ideal RNA world. And there really are deep distinctions between archaea and bacteria, distinctions that can't be explained away by some sleight of hand. They certainly imply that life started out with a retroviral lifecycle.

I'm genuinely excited that the story we've unravelled here might just be true, but there is one deep uncertainty at the back of my mind: the implication that cellular life emerged twice from the deep-ocean vents. Did swarming RNAs infect nearby vents, eventually taking over great swathes of the ocean, enabling selection to operate on a global scale? Or was there something uniquely favourable about one particular vent system, whose atypical conditions gave rise to both the archaea and bacteria? Perhaps we'll never know; but the play of chance and necessity should give everybody some pause for thought.

PHOTOSYNTHESIS

Summoned by the Sun

Imagine a world without photosynthesis. It wouldn't be green, for a start. Our emerald planet reflects the glory of plants and algae, and ultimately their green pigments, which absorb light for photosynthesis. First among pigments is the marvellous transducer that is chlorophyll, which steals a beam of light and conjures it into a quantum of chemical energy, driving the lives of both plants and animals.

The world probably wouldn't be blue either, for the azures of the heavens and the marines of the oceans depend on clear skies and waters, cleansed of their haze and dust by the scouring power of oxygen. And without photosynthesis there would be no free oxygen.

In fact there might not be any oceans either. Without oxygen there is no ozone; and without that, there is little to cut down the searing intensity of ultraviolet rays. These split water into oxygen and hydrogen. The oxygen is formed slowly and never builds up in the air; instead it reacts with iron in the rocks, turning them a rusty-red colour. And hydrogen, the lightest of gases, evades the tug of gravity and slips away into space. The process may be slow but it is also inexorable: the oceans bleed into space. Ultraviolet radiation cost Venus its oceans, and maybe Mars too.

So we don't need much imagination to picture a world without photosynthesis: it would look a lot like Mars, a red dusty place, without oceans, and without any overt signs of life. Of course, there is life without photosynthesis, and many astrobiologists seek it on Mars. But even if a few bacteria are found hiding beneath the surface, or buried in an icecap, the planet itself is dead. It is in near-perfect equilibrium, a sure sign of inertia. It could never be mistaken for Gaia.

Oxygen is the key to planetary life. No more than a waste product of photosynthesis, oxygen really is the molecule that makes a world. It is let loose by photosynthesis so fast that it finally overwhelms the capacity of a planet to swallow it up. In the end, all the dust and all the iron in the rocks, all the sulphur in the seas and methane in the air, anything that can be oxidised is oxidised, and free oxygen pours into the air and the oceans. Once there, oxygen puts a stop to the loss of water from the planet. Hydrogen, when released from water, inevitably bumps into more oxygen before it finds its way out into space. Swiftly it reacts to form water again, which now rains back down from the heavens, drawing to a halt the loss of the oceans. And when oxygen accumulates in the air, an ozone shield forms, ablating the searing intensity of the ultraviolet rays, and making the world a more habitable place.

Oxygen doesn't just rescue a planet's life: it energises all life, and makes it big. Bacteria can do perfectly well without oxygen: they have an unparalleled skill at electrochemistry, they are able to react together virtually all molecules to glean a little energy. But the sum total of energy that can be derived from fermentation, or by reacting two molecules like methane and sulphate together, is negligible in comparison with the power of oxygen respiration – literally the burning up of food with oxygen, oxidising it fully to carbon dioxide and water vapour. Nothing else can provide the energy needed to fuel the demands of multicellular life. All animals, all plants, all of them depend on oxygen for at least part of their life cycle. The only exception that I'm aware of is a microscopic (but multicellular) nematode worm that somehow gets along in the stagnant oxygen-free depths of the Black Sea. So a world without free oxygen is microscopic, at least at the level of individual organisms.

Oxygen contributes to large size in other ways too. Think of a food chain. The top predators eat smaller animals, which might in turn eat insects, which eat smaller insects, which live on fungus or leaves. Five or six levels in a food web are not uncommon. At each step energy is wasted, for no form of respiration is ever 100 per cent efficient. In fact, oxygen respiration is about 40 per cent efficient, while most other forms of respiration (using iron or sulphur instead of oxygen, for example) are less than 10 per cent efficient. This means that, without using oxygen, the energy available dwindles to 1 per cent of the initial input in only two levels, whereas with oxygen it takes six levels to arrive at the same point. That in turn means that long food chains are only feasible with oxygen respiration. The economy of the food chain means that predators can operate in an oxygenated world, but predation as a lifestyle just doesn't pay without oxygen.

Predation escalates size, of course, driving arms races between predator and prey. Shells combat teeth, camouflage tricks the eye; and size intimidates both hunter and hunted. With oxygen, then, predation pays; and with predators size pays. So oxygen makes large organisms not just feasible but also probable.

It also helps build them. The protein that gives animals their tensile strength is collagen. This is the main protein of all connective tissues, whether calcified in bones, teeth and shells, or 'naked' in ligaments, tendons, cartilage and skin. Collagen is by far the most abundant protein in mammals, making up a remarkable 25 per cent of total body protein. Outside the vertebrates, it is also the critical component of shells, cuticles, carapaces and fibrous tissues of all sorts – the 'tape and glue' of the whole animal world. Collagen is composed of some unusual building blocks, which require free oxygen to form cross-links between adjacent protein fibres, giving the overall structure a high tensile strength. The requirement for free oxygen means that large animals, protected with shells or strong skeletons, could only evolve when atmospheric oxygen levels were high enough to support collagen production – a factor that might have contributed to the abrupt appearance of large animals in the fossil record at the beginning of the Cambrian period, some 550 million years ago, soon after a big global rise in atmospheric oxygen.

The need for oxygen to make collagen may seem no more than an accident; if not collagen why not something else with no requirement for free oxygen? Is oxygen necessary to give strength or just a random ingredient that happened to be incorporated and then forever remained part of the recipe? We don't really know, but it's striking that higher plants, too, need free oxygen to form their structural support, in the shape of the immensely strong polymer lignin, which gives wood its flexible strength. Lignin is formed in a chemically haphazard way, using free oxygen to form strong cross-links between chains. These are very difficult to break down, which is why wood is so strong and why it takes so long to rot. Eliminate lignin from trees – a trick that manufacturers of paper have tried, as they need to remove it laboriously from wood pulp to make paper – and the trees slump to the ground, unable to sustain their own weight even in the lightest breeze.

So without oxygen there would be no large animals or plants, no predation, no blue sky, perhaps no oceans, probably nothing but dust and bacteria. Oxygen is without a doubt the most precious waste imaginable. Yet not only is it a waste product, it is also an unlikely one. It is quite feasible that photosynthesis could have evolved here on earth, or Mars, or anywhere else in the universe, without ever producing any free oxygen at all. That would almost certainly consign any life to a bacterial level of complexity, leaving us alone as sentient beings in a universe of bacteria.

One reason why oxygen might never have accumulated in the air is respiration. Photosynthesis and respiration are equal and opposite processes. In a nutshell, photosynthesis makes organic molecules from two simple molecules, carbon dioxide and water, using sunlight to provide the energy needed. Respiration does exactly the opposite. When we burn organic molecules (food) we release carbon dioxide and water back into the air; and the energy released is what powers our lives. All our energy is a beam of sunlight set free from its captive state in food.

Photosynthesis and respiration oppose each other not just in the details of their chemistry, but also in global accounting. If there was no respiration – no animals, fungi and bacteria burning up plant food – then all the carbon dioxide would have been sucked out of the atmosphere long ago, converted into biomass. Everything would then more or less grind to a halt, bar the trickle of carbon dioxide set free by slow decay or volcanoes. But this is far from what really happens. What really happens is that respiration burns all the organic molecules put away by plants: on a geological timeframe, plants disappear in a puff of smoke. This has one profound consequence. All the oxygen put in the air by photosynthesis is taken out again by respiration. There is a long-term, unchanging, never-ending equilibrium, the kiss of death for any planet. The only way that a planet can gain an oxygen atmosphere – the only way it can escape the dusty red fate of Mars – is if a little plant matter is preserved intact, immune to the elements and to life's ingenuity in finding ways of breaking it down for energy. It must be buried.

And so it is. Preserved plant matter is buried as coal, oil, natural gas, soot, charcoal or dust, in rocks deep in the bowels of the earth. According to the ground-breaking geochemist Robert Berner, recently retired from Yale, there is around 26,000 times more 'dead' organic carbon trapped in the earth's crust than in the entire living biosphere. Each atom of carbon is the antithesis of a molecule of oxygen in the air. For every atom of carbon that we dig up and burn as fossil fuel, a molecule of oxygen is stripped out of the air, and converted back to carbon dioxide, with serious, albeit unpredictable, consequences for climate. Luckily we will never deplete the world's oxygen supply by burning fossil fuels, even if we do play havoc with the climate: the vast majority of organic carbon is buried as microscopic detritus in rocks like shales, inaccessible to human industry, or at least economic industry. So far, despite our vainglorious efforts to burn all the known reserves of fossil fuels, we have lowered the oxygen content of the air by a mere two or three parts per million, or about 0.001 per cent.[1]

But this vast reservoir of buried organic carbon is not formed continuously – it has been buried in fits and starts over the geological aeons. The norm is very close to an exact balance, in which respiration cancels out photosynthesis (and erosion cancels out any burial), so there is next to no net burial. This is why oxygen levels have remained at around 21 per cent for tens of millions of years. In deep geological time, though, on rare occasions, things were very different. Perhaps the most striking example is the Carboniferous period, 300 million years ago, when dragonflies as big as seagulls flapped through the air and millipedes a metre long crawled the undergrowth. These giants owed their very existence to the exceptional rate of carbon burial in Carboniferous times, whose huge coal reserves give the era its name. As carbon was buried beneath the coal swamps, oxygen levels soared above 30 per cent, giving some creatures an opportunity to grow far beyond their normal bounds of size – specifically, animals that rely on the passive diffusion of gases down tubes or across the skin, like dragonflies, rather than the active ventilation of lungs.[2]

What was behind the unprecedented rate of carbon burial in Carboniferous times? A variety of accidental factors, almost certainly. The alignment of the continents, the wet climate, the great flood plains; and perhaps most importantly the evolution of lignin, which gave rise to large trees and sturdy plants capable of colonising large areas of the landmass. Lignin, tough to break down for fungi and bacteria even today, seems to have been an insurmountable challenge soon after its evolution. Rather than being broken down for energy, it was buried intact on a vast scale, and its antithesis, oxygen, flooded the air.

Geological accidents colluded on two other occasions to force up oxygen levels, both perhaps the outcome of global glaciations called 'snowball earths'. The first great rise in oxygen levels, around 2,200 million years ago, followed hard on the heels of a period of geological upheavals and global glaciation around that time; and a second period of global glaciations, around 800 to 600 million years ago, also seems to have pushed up oxygen levels. Such calamitous global events probably altered the balance of photosynthesis to respiration, and of burial to erosion. As the great glaciers melted and the rains fell, minerals and nutrients (iron, nitrates and phosphates) that had been scoured from the rocks by ice were washed into the oceans, causing a great bloom of photosynthetic algae and bacteria, similar to, but far greater than, those caused today by fertilisers. Not only would such a run-off induce a bloom, it would also tend to bury it: the dust, dirty ice and grit washed into the oceans mixed with blooming bacteria and settled out, burying carbon on an unprecedented scale. And with it came a lasting global rise in oxygen.

So there is a sense of the accidental about the oxygenation of our planet. This sense is reinforced by the absence of change for long periods otherwise. From 2,000 million to around 1,000 million years ago – a period geologists call the 'boring billion' – almost nothing of note seems to have happened. Oxygen levels remained steady and low throughout this period, as indeed they did at other times for hundreds of millions of years. Stasis is the default, while episodes of geological restlessness wreak lasting change. Such geological factors might intervene on other planets too; but tectonic movements and active volcanism seem to be necessary to bring about the accidental conjugations needed for oxygen to accumulate. It is not beyond the bounds of possibility that photosynthesis evolved long ago on Mars, but that this small planet, with its shrinking volcanic core, could not sustain the geological flux required for oxygen to accumulate, and later expired on a planetary scale.

But there is a second and more important reason why photosynthesis need not lead to an oxygen atmosphere on a planet. Photosynthesis itself may never turn upon water as a raw material. We are all familiar with the form of photosynthesis that we see around us. Grasses, trees, seaweeds, all operate in fundamentally the same way to release oxygen – a process known as 'oxygenic' photosynthesis. But if we take several steps back and consider bacteria, there are many other options. Some relatively primitive bacteria make use of dissolved iron or hydrogen sulphide instead of water. If these sound like implausible raw materials to us, it is only because we have become so inured to our oxygenated world – the product of 'oxygenic' photosynthesis – that we struggle to imagine conditions on the early earth when photosynthesis first evolved.

We also struggle to grasp the counterintuitive, but in fact simple, mechanism of photosynthesis. Let me give an example, which I suspect, perhaps unfairly, illustrates the general perception of photosynthesis. This is Primo Levi, from his lovely book *The Periodic Table*, published in 1975 and voted the 'best popular science book ever' by an audience (including me) at the Royal Institution in London, in 2006:

Our atom of carbon enters the leaf, colliding with other innumerable (but here useless) molecules of nitrogen and oxygen. It adheres to a large and complicated molecule that activates it, and simultaneously receives the decisive message from the sky, in the flashing form of a packet of solar light: in an instant, like an insect caught by a spider, it is separated from its oxygen, combined with hydrogen and (one thinks) phosphorus, and finally inserted in a chain, whether long or short does not matter, but it is the chain of life.

Spot the mistake? There are actually two, and Levi ought to have known better, for the true chemistry of photosynthesis had been elucidated forty years earlier. A flashing packet of solar light does not activate carbon dioxide: it can be activated just as well in the middle of the night, and indeed is never activated by light, even in the brightest sunshine. Nor is carbon separated in an instant from its oxygen. Oxygen remains stubbornly bound to its carbon. Underpinning Levi's account is the common, but plain wrong, assumption that the oxygen released by photosynthesis comes from carbon dioxide. It does not. It comes from water. And that makes all the difference in the world. It is the first step to understanding how photosynthesis evolved. It is also the first step to solving the energy and climate crises of our planet.

The packets of solar energy used in photosynthesis split water into hydrogen and oxygen: the same reaction that occurs on a planetary scale when the oceans bleed into space, driven away by the blast of ultraviolet radiation. What photosynthesis achieves – and what we have so far failed to achieve – is to come up with a catalyst that can strip the hydrogen from water with a minimal input of energy, using gentle sunlight rather than searing ultraviolet or cosmic rays. So far, all our human ingenuity ends up consuming more energy in splitting water than is gained by the split. When we succeed in mimicking photosynthesis, with a simple catalyst that gently prises hydrogen atoms from water, then we will have solved the world's energy crisis. Burning that hydrogen would comfortably supply all the world's energy needs, and regenerate water as the only waste: no pollution, no carbon footprint, no global warming. Yet this is no easy task, for water is a marvellously stable combination of atoms, as the oceans attest; even the most furious storms, battering cliffs, don't break water into its component atoms. Water is at once the most ubiquitous and unattainable raw material on our planet. The modern mariner might muse on how to power his boat with water and a splash of sunshine. He should ask the green scum floating on the waves.

The same problem, of course, faced the remote ancestors of that scum, the ancestors of today's cyanobacteria, the only form of life on our planet to have chanced upon the trick of splitting water. The strange thing is that cyanobacteria split water for exactly the same reason that their bacterial relatives split hydrogen sulphide or oxidise iron: they want the electrons. And on the face of it, water is the last place to find them.

Photosynthesis is conceptually simple: it's all about electrons. Add a few electrons to carbon dioxide, along with a few protons to balance out charges, and, hey presto, there you have it – a sugar. Sugars are organic molecules: they are Primo Levi's chain of life and the ultimate source of all our food. But where do the electrons come from? With a little energy from the sun they can come from more or less anywhere. In the case of the familiar 'oxygenic' form of photosynthesis, they come from water; but in fact it's far easier to strip them from other compounds less stable than water. Take electrons from hydrogen sulphide and instead of releasing oxygen into the air you deposit elemental sulphur – biblical brimstone. Take them from iron dissolved in the oceans (as ferrous iron) and you get rusty-red ferric iron, which settles out as new rocks – a process that might once have been responsible for the vast 'banded-iron formations' found around the world, and today the largest remaining reserves of low-grade iron ore.

These forms of photosynthesis are marginal in today's oxygen-rich world, simply because the raw materials, hydrogen sulphide or dissolved iron, are rarely found in sunny well-aerated waters. But when the earth was young, before the rise of free oxygen, they would have been by far the easiest source of electrons, and they saturated the oceans. This raises a paradox, whose resolution is fundamental to understanding how photosynthesis first evolved. Why switch from a plentiful and comfortable source of electrons to something far more problematic, water, whose waste product, oxygen, was a toxic gas capable of causing grievous bodily harm to any bacteria that produced it? The fact that, given the power of the sun and a clever catalyst, water is far more abundant than either, is beside the point, for evolution has no foresight. So too is the fact that oxygenic photosynthesis transformed the world; the world cares not a whit. So what kind of environmental pressure, or mutations, could have driven such a shift?

The facile answer, which you'll find in plenty of textbooks, is that the raw materials ran out: life turned to water because there were no easy alternatives left, just as we might turn to water when we run out of fossil fuels. But this answer can't be true: the geological record makes it plain that 'oxygenic' photosynthesis evolved long before – more than a billion years before – all these raw materials ran out. Life was not forced into a corner.

Another answer, only now emerging, lies hidden in the machinery of photosynthesis itself, and is altogether more beautiful. It is an answer that combines chance and necessity, an answer that shines the light of simplicity on one of the most convoluted and complicated extractions in the world.

In plants the business of electron extraction takes place in the chloroplasts. These are the minute green structures found in the cells of all leaves, all blades of grass, imparting their own green to the leaves as a whole. The chloroplasts are named from the pigment that gives them their colour in turn. This is chlorophyll, which is responsible for absorbing the energy of the sun in photosynthesis. Chlorophyll is embedded in an extraordinary membrane system that makes up the inside of the chloroplasts. Great stacks of flattened disks, looking for all the world like an alien power station in a science fiction movie, are connected to each other via flying tubes, which criss-cross the vertiginous spaces at all angles and heights. In the disks themselves, the great work of photosynthesis takes place: the extraction of electrons from water.

If the extraction of electrons from water is difficult, plants make an extraordinary meal of it. The complexes of proteins and pigments are so vast, in molecular terms, that they amount to a small city. Altogether they form into two great complexes, known as Photosystem I and Photosystem II, and each chloroplast contains thousands of such photosystems. Their job is to catch a beam of light, and transform it into living matter. Working out how they do it has taken the best part of a century, and took some of the most elegant and ingenious experiments ever done. This, sadly, is not the place to discuss them.[3] Here we need concern ourselves only with what we have learned, and what that has to say about the invention of photosynthesis.

The conceptual heart of photosynthesis, the roadmap that makes sense of it all, is known as the 'Z scheme', a formulation that fascinates and horrifies biochemistry students in equal measure. First laid out by the brilliant but diffident Englishman Robin Hill in 1960, the Z scheme describes the 'energy profile' of photosynthesis. Hill's utterances were notoriously gnomic. Wishing not to cause offence by labouring the obvious, even those in his own lab were surprised when his hypothesis appeared in *Nature* in 1960, having had little idea what he was working on. In fact the Z scheme was not based on Hill's own work, which had been of primary importance, but rather was distilled from a number of puzzling experimental observations. Foremost among these was a curious matter of thermodynamics. Photosynthesis, it turned out, produced not just new organic matter but also ATP, the 'energy currency' of life. Quite unexpectedly, these two seemed always to be coupled: the more organic matter produced by photosynthesis, the more ATP, and vice versa (if the amount of organic matter falls, so too does ATP production). The sun apparently provides two free lunches in synchrony. Robin Hill, remarkably, had the insight to grasp the whole mechanism of photosynthesis from this single fact. Genius, it is said, is the ability to see the obvious before anyone else.[4]

And yet – typically for anything associated with Hill – even the term 'Z scheme' is gnomically misleading. The Z should really be rotated through 90 degrees to become an 'N'; then it would reflect more accurately the energy profile of photosynthesis. Picture the first upstroke of the 'N' as a vertical uphill reaction: energy must be supplied to make it work. The diagonal downstroke of the 'N' is then a downhill reaction – it releases energy that can be captured and stored in the form of ATP. The final upstroke is again an uphill reaction, which requires an input of energy.

In photosynthesis, the two photosystems – Photosystem I and II – lie at the two bottom points of the 'N'. A photon of light hits the first photosystem and blasts an electron up to a higher energy level; the energy of this electron then cascades down in a series of small molecular steps, which provide the energy needed to make ATP. Back at a low energy level, the electron arrives at the second photosystem, where a second photon blasts it up a second time to a higher energy level. From this second high point, the electron is ultimately

Figure 3.1 Cartoon depicting the Z scheme, by Richard Walker. The energy of a photon, depicted as a mallet blow, blasts an electron to a high energy level. As the electron cascades back down to a lower energy level, some of the energy released powers work in the cell. A second photon then blasts the electron back up to an even higher energy level, where it is captured in the form of a high-energy molecule (NADPH) that later reacts with carbon dioxide to form an organic molecule.

transferred to carbon dioxide, in the first step of making a sugar. One helpful cartoon by Richard Walker (see Fig. 3.1) depicts the process as a fairground test-of-strength game, where a punter hits a pallet with a mallet to ring a bell by forcing a metal ringer up a pole. In this case, the swing of the mallet provides the energy to blast the ringer up the pole; in the case of photosynthesis, the energy of a photon from the sun does the same job.

The Z scheme, or N scheme if you prefer, is a curiously convoluted way of going about things, but there are good technical reasons for it. It verges on the chemically impossible to couple the removal of electrons from water to the conversion of carbon dioxide into a sugar in any other way. The reason relates to the nature of electron transfer, and specifically the chemical affinity of electrons for particular compounds. Water is very stable, as we've seen: it has a high affinity for its electrons. To steal an electron from water requires huge pulling power, which is to say a very powerful oxidant. That powerful oxidant is a voracious form of chlorophyll, a molecular Mr Hyde, transformed from the meek Dr Jekyll by absorbing photons of high energy.[5] But an entity that is good at pulling tends to be less good at pushing. A molecule that grasps electrons tightly is chemically disinclined to push them away, just as the misanthropic Mr Hyde, or any grasping miser, is not prone to give away his wealth in acts of spontaneous generosity. So it is with this form of chlorophyll. When activated by light it has tremendous power to pull electrons from water, but little strength to push them away anywhere else. In the jargon, it is a powerful oxidant but a weak reductant.

Carbon dioxide raises the reverse problem. It is also very stable, and has no chemical desire to be stuffed with more electrons. It will only grudgingly accept electrons from a pusher of great strength – in the jargon, a strong reductant. This requires a different form of chlorophyll: one that is very good at pushing, and poor at pulling. Rather than being a grasping miser, it is more like a street hustler, intent on forcing dodgy goods onto vulnerable passers-by. When activated by light, this form of chlorophyll has the power to force its electrons onto another molecule that wants equally to be rid of them, its conspirator in crime and co-hustler, NADPH, and ultimately onto carbon dioxide.[6]

So there is a reason for having two photosystems in photosynthesis. No real surprise there. But the more challenging question is: how did such a complex interrelated system come to evolve? There are actually five parts to this sequence. First is the 'oxygen-evolving complex', a kind of molecular nutcracker that positions the water molecules just so to have their electrons cracked out one by one, releasing oxygen as waste. Then comes Photosystem II (rather confusingly, the two photosystems are named in reverse order, for historical reasons), which when activated by light transforms into the molecular Mr Hyde, and yanks out these electrons from the oxygen-evolving complex. Then comes an electron-transport chain, which transfers the electrons away, like rugby players passing a ball across a pitch. The electron-transport chain uses the downhill energy gradient to make a little ATP, before delivering up the same electrons to Photosystem I. Here another photon blasts them up to a high energy level again, where they are held in trust by the molecular 'hustler' NADPH, a strong pusher of electrons, which wants nothing better than to be rid of them again. And then finally comes the molecular machinery needed to activate carbon dioxide and convert it to a sugar. Using the molecular hustler generated by Photosystem I, the conversion of carbon dioxide into a sugar is powered by chemistry rather than light, and is actually known as the dark reaction – a feature that Primo Levi failed to appreciate.

These five systems work in sequence to strip electrons from water and push them onto carbon dioxide. It's an enormously complicated way to crack a nut, but it seems to be about the only way to crack this particular nut. The great evolutionary question is how did all these complex interrelated systems come into existence, and come to be organised in exactly the right way, perhaps the only way, to make oxygenic photosynthesis work?

The word 'fact' is always likely to make biologists tremble in their boots, as there are so many exceptions to every rule; but one such 'fact' is virtually certain about oxygenic photosynthesis – it only evolved once. The seat of photosynthesis, the chloroplast, is found in all photosynthetic cells of all plants and all algae. Chloroplasts are omnipresent and are obviously related to each other. They share a secret history. The clue to their past lies in their size and shape: they look like little bacteria living inside a larger host cell (see Fig. 3.2). This hint of bacterial ancestry is confirmed by the existence of independent rings of DNA in all chloroplasts. These rings of DNA are copied whenever chloroplasts divide, and passed on to the daughters in the same way as bacteria. The detailed sequence of letters in chloroplast DNA not only corroborates the link with bacteria, but also points an accusing finger at the closest living relative: cyanobacteria. Last but not least, the Z scheme of plant photosynthesis, along with all five of its component parts, is presaged exactly (if with simpler machinery) in cyanobacteria. In short, there is no doubt that chloroplasts were once free-living cyanobacteria.

Once misnamed, poetically, the 'blue-green algae', the cyanobacteria are the only known group of bacteria that can split water via the 'oxygenic' form

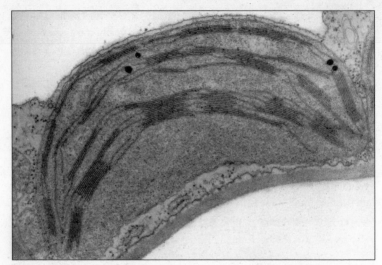

Figure 3.2 Classic view of a chloroplast from beet (*Beta vulgaris*), showing the stacks of membranes (thylakoids) where water is split apart to release oxygen in photosynthesis. The resemblance to a bacterium is not accidental: chloroplasts were once free cyanobacteria.

of photosynthesis. Exactly how some of their number came to live within a larger host cell is a mystery wrapped in the shrouds of deep geological time. It undoubtedly happened more than 1,000 million years ago, but presumably they were simply engulfed one day, survived digestion (not uncommon), and ultimately proved useful to their host cell. The host, impregnated with cyanobacteria, went on to found two great empires, the algae and the plants, for all of them today are defined by their ability to live on sun and water, by way of the photosynthetic apparatus inherited from their bacterial guests.

So the quest for the origin of photosynthesis becomes the quest for the origin of cyanobacteria, the only type of bacteria to have cracked the problem of splitting water. And this is one of the most controversial, and indeed still unresolved, stories in modern biology.

Until the turn of the millennium, most researchers were persuaded, if vexed, by the remarkable findings of Bill Schopf, an energetic and combative

Figure 3.3 Living stromatolites in Hamelin Pool, near Shark Bay, western Australia. The pool is approximately double the salinity of the open ocean, which stifles grazers like snails and enables the cyanobacterial colonies to flourish.

professor of palaeobiology at the University of California, Los Angeles. From the 1980s, Schopf had discovered and analysed a number of the oldest fossils of life on earth – some 3,500 million years old. The word 'fossil' needs a little clarification here. What Schopf found were strings of microscopic rock capsules, which looked a lot like bacteria, and were about the right size. From their detailed structure, Schopf initially proclaimed the fossils to be cyanobacteria. These tiny microfossils were often associated with what looked like fossil stromatolites. Living stromatolites are mineralising domes that grow in incremental layers, up to a metre or so in height, which are formed by thriving communities of bacteria that encrust the buried mineral layers (see Fig. 3.3). Eventually the entire structure turns to solid rock, often strikingly beautiful in section. The outer, living layers of modern stromatolites are usually heaving with cyanobacteria, so Schopf was able to claim these ancient forms as further evidence for the early appearance of cyanobacteria. Lest there be any doubt, Schopf went on to show that these putative fossils contained remnants of organic carbon, of a sort that seemed diagnostic of life – and not just any old life, but photosynthetic life. All in all, said Schopf, cyanobacteria, or something that looked very much like them, had already evolved on the earth by 3,500 million years ago, just a few hundred million years after the end of the great asteroid bombardment that marked the earliest years of our planet, so soon after the formation of the solar system itself.

Few people were equipped to challenge Schopf's interpretation of these ancient fossils, and the few that were also seemed convinced. Others, though, if less expert, were more sceptical. It was not easy to reconcile the early evolution of cyanobacteria – presumably belching out oxygen as a waste product, as they do today – with the first geological signs of oxygen in the atmosphere, well over a billion years later. And perhaps more seriously still, the complexity of the Z scheme made most biologists baulk at the idea that oxygenic photosynthesis could have evolved so fast. The other forms of photosynthesis, being simpler, seemed more in keeping with great antiquity. Overall, then, most people accepted that these were bacteria, perhaps photosynthetic bacteria, but there were doubts about whether they were really cyanobacteria, the pinnacle of the art.

Then Martin Brasier, professor of palaeobiology at Oxford, stepped into the ring, in what turned into one of the great fights of modern paleontology, a science in any case noted for the passion of its protagonists, and the elasticity of much of its evidence. Most researchers interested in the early fossils relied on specimens deposited in the Natural History Museum, London, but Brasier returned to the geological setting where Schopf had originally dug up his fossils, and expressed shock. Far from being the shallow, tranquil seafloor posited by Schopf, the entire region was shot through with geothermal veins, evidence of a tumultuous geological past, said Brasier. Schopf had hand-picked his specimens to make his case, Brasier went on, and had concealed other specimens, superficially similar, but patently not biological; they were probably all formed by the action of scalding water on mineral sediments. The stromatolites too, he said, were formed by geological processes, not bacteria, and were no more mysterious than ripples in the sand. And the organic carbon had no microscopic structure at all, making it quite indistinguishable from the inorganic graphite found in many geothermal settings. Finally, as if to drive a stake through the corpse of a once-great scientist, one former graduate student recalled being bullied and forced into making dubious interpretations. Schopf looked like a broken man.

But never a man to take a beating lightly, Schopf emerged fighting. Assembling more data to make his case, he met Brasier onstage at a fiery NASA spring meeting in April 2002, and the pair defended their corners. Brasier, every inch the haughty Oxford don, condemned Schopf's case as 'a truly hydrothermal performance – all heat and not much light'. Even so, the jury has not really been convinced by either side. While there is real doubt about the biological origin of the earliest microfossils, others, dating to only a hundred million years later, are less contested; and Brasier himself has put forward candidate fossils from this time. Most scientists, including Schopf, now apply more stringent criteria to verify biological provenance. The one casualty so far is the cyanobacteria, once the centrepiece of Schopf's fame. Even Schopf concedes that the microfossils are probably not cyanobacteria, or at least are no more likely to be cyanobacteria than any other type of filamentous bacteria. And so we have arrived back in the starting blocks, chastened, and with no better idea of the evolution of cyanobacteria than we had at the outset.

I use this tale to illustrate just how difficult it is to fathom the depths of geological time using the fossil record alone. Even proving the existence of cyanobacteria, or at least their ancestors, does not prove that they had already stumbled upon the means of splitting water. Perhaps their ancestors relied on a more primitive form of photosynthesis. But there are other ways to mine information from deep time that may yet prove more informative. These are the secrets buried within living things themselves, both in their genes and in their physical structures, especially their protein structures.

Over the last two or three decades, the detailed molecular structures of both plant and bacterial photosystems have come under intense scrutiny, with scientists bringing to bear a great battery of techniques with daunting names, signifying no less daunting methodologies, from X-ray crystallography to electron-spin resonance spectroscopy. How these techniques work needn't concern us here; suffice to know that they have been used to map out the shapes and structures of the photosynthetic complexes in nearly, but tantalisingly not quite, atomic resolution. Even now, arguments rage at meetings, but they are arguments about details. As I write, I have recently returned from a discussion meeting at the Royal Society in London that was rich in argument about the exact location of five critical atoms in the oxygen-evolving complex. The arguments were at once fiddling and profound. Profound, because their exact position defines the strict chemical mechanism by which water is split; knowing this is the key step to solving the world's energy crisis. But fiddling, because their squabbles are about positioning these five atoms to within a space of a few diameters of an atom – a few angstroms (less than a millionth of a millimetre). To the astonishment of the older generation of researchers, there is little real disagreement about the position of all the other 46,630 atoms of Photosystem II, mapped out by Jim Barber's team at Imperial College in 2004, and more recently in even more detail.

While these few atoms have yet to be allocated their final resting places, the larger architecture of the photosystems, hinted at for more than a decade, is now plain, and speaks volumes about their evolutionary history. In 2006, a small team led by Bob Blankenship, now a distinguished professor at the University of Washington in St Louis, showed that the two photosystems are extraordinarily well conserved in bacteria.[7] Despite enormous evolutionary distances between the various groups of bacteria, the core structures of the photosystems are almost identical, to the point that they can be superimposed in space using a computer. In addition, Blankenship confirmed another link that researchers had suspected for a long time: the two photosystems (Photosystem I and II) also share core structures, and almost certainly evolved from a common ancestor, long, long ago.

In other words, there was once a single photosystem. At some point the gene became duplicated to give two identical photosystems. These slowly diverged from each other under the influence of natural selection, while retaining a close structural similarity. Ultimately, the two photosystems were yoked together in the Z scheme of cyanobacteria, and later passed on to the plants and algae in chloroplasts. But this simple narrative conceals a fascinating dilemma. Duplicating a primitive photosystem could never solve the problem of oxygenic photosynthesis – it could never couple a strong puller to a strong pusher of electrons. Before photosynthesis could work, the two photosystems had to diverge in opposite directions, and only then could they become usefully interlinked. So the question is, what succession of events could drive them apart, only to link them together again as intimate but opposite partners, like man and woman, reunited after diverging from an egg?

The best way to find the answer is to look to the photosystems themselves. These are united in the Z scheme of cyanobacteria, but otherwise have interestingly deviant evolutionary histories. Let's put aside for the moment where the photosystems originally came from, and take a quick look at their current distribution in the bacterial world. Apart from the cyanobacteria, they are never found together in the same bacterium. Some groups of bacteria have only Photosystem I, while other groups have only Photosystem II. Each photosystem works by itself to achieve different ends; and their precise tasks give a striking insight into how oxygenic photosynthesis first evolved.

In bacteria, Photosystem I does exactly the same as it does in plants. It draws electrons from an inorganic source and forms a molecular 'street hustler' that pushes them on to carbon dioxide to make sugars. What differs is the inorganic source of electrons. Rather than water, which it can't handle at all, Photosystem I draws electrons from hydrogen sulphide or iron, both of which are far easier targets than water. Incidentally, the molecular 'hustler' formed by Photosystem I, NADPH, can also be formed by pure chemistry, for example in the hydrothermal vents we discussed in Chapter 1. Here too NADPH is used to convert carbon dioxide into sugars via a similar set of reactions. So the only real innovation of Photosystem I was to harness light to do a job that was previously done by chemistry alone.

It's also worth noting here that there is nothing particularly special about the capacity to convert light into chemistry: almost any pigment can do it. The chemical bonds in pigments are good at absorbing photons of light. When they do so an electron is zapped up to a higher energy level, and nearby molecules can easily capture it. As a result the pigment becomes photo-oxidised:

it is in need of an electron to balance the books, and takes one from iron or hydrogen sulphide. This is all that chlorophyll does. Chlorophyll is a porphyrin, not dissimilar in its structure to haem, the pigment that carries oxygen in our blood. Many other porphyrins can pull off similar tricks with light, sometimes with unwelcome consequences, as in diseases like porphyria.[8] And crucially, porphyrins are among the more complex molecules that have been isolated from asteroids and synthesised in the lab under plausibly prebiotic conditions. Porphyrins, in other words, would most likely have formed spontaneously on the early earth.

In short, Photosystem I took a simple-enough pigment, a porphyrin, and coupled its spontaneous light-driven chemistry to reactions that take place in bacterial cells anyway. The outcome was a primitive form of photosynthesis that could use light to strip electrons from 'easy' sources, such as iron and hydrogen sulphide, and pass the electrons on to carbon dioxide to form sugars. Thus these bacteria use light to make food.

What about Photosystem II? The bacteria that use this photosystem use light to pull off quite a different trick. This form of photosynthesis doesn't produce organic matter. Rather, it converts light energy into chemical energy, indeed electricity, which can be used to power the cell. The mechanism is very simple. When a photon strikes a molecule of chlorophyll, one electron is zapped up to a higher energy level, as before, where it is captured by a nearby molecule. This electron is then passed hot-handed from carrier to carrier down an electron-transport chain, each time releasing a little energy, until it has returned to a low energy level. Some of the energy released in this process is captured, to make ATP. Finally, the exhausted electron is returned to the same chlorophyll that it started out from, completing the circuit. In short, light zaps an electron to a high energy level, and, as it cascades back down to a 'resting' level, the release of energy is captured as ATP, a form of energy that the cell can use. It's just a light-powered electric circuit.

How did such a circuit come to be? Again, the answer is by mixing and matching. The electron-transport chain is more or less the same as that used for respiration, which evolved in the vents as we saw in Chapter 1; it was just borrowed for an ever-so-slightly new purpose. In respiration, as we noted, electrons are stripped from food and passed, ultimately, to oxygen, to form water. The energy released is used to generate ATP. In this form of photosynthesis, exactly the same thing happens: high-energy electrons are passed along a chain, not to oxygen, but to a 'grasping' (oxidising) form of chlorophyll. The more that the chlorophyll can 'pull' electrons (that is, the closer it is to oxygen in chemical character), the more efficient the chain will be, sucking electrons along and drawing out their energy. The great advantage is that no fuel, or food, is needed, at least to provide energy (it is needed to synthesise new organic molecules).

As a general conclusion, then, the simpler forms of photosynthesis are mosaic-like in character. Both forms plugged a new transducer, chlorophyll, into existing molecular machinery. In one case, this machinery converts carbon dioxide into sugars; in the other, it produces ATP. As to chlorophyll, similar porphyrin pigments probably formed spontaneously on the early earth, and natural selection did the rest. In each case, small changes in the structure of chlorophyll alter the wavelength of light absorbed, and so the chemical properties. All these changes alter the efficiency of processes that happen spontaneously, albeit initially much more wastefully. The natural outcome is to produce a 'grasping miser' form of chlorophyll for ATP synthesis in some footloose bacteria and a 'street hustler' type of chlorophyll to make sugars in bacteria living close to supplies of hydrogen sulphide or iron. But we're still left with the bigger question: how did it all come to be tied together in the Z scheme of cyanobacteria, to split the ultimate fuel, water?

The short answer is we don't know for sure. There are ways of finding a definite answer, but unfortunately they haven't worked. For example, we can systematically compare and contrast the genes for the photosystems in bacteria, to build a gene tree that betrays the ancestry of the photosystems. Such trees are felled, though, by a fact of bacterial life – sex. Bacterial sex is not like our own, in which genes are inherited down the generations, giving rise to a nicely ordered family tree. Bacteria throw their genes around with a profligate disregard for the labours of geneticists. The result is more of a web than a tree, in which the genes of some bacteria end up in other, totally unrelated bacteria.

And that means we have no real genetic evidence for how the photosystems came to be assembled together in the Z scheme.

But that doesn't mean we can't work out the answer. The great value of hypotheses in science is that, by making imaginative leaps into the unknown, they suggest new angles and experiments that can corroborate or refute the postulates. Here is one of the best – a beautiful idea from John Allen, professor of biochemistry at Queen Mary, University of London, and an inventive mind. Allen has the dubious distinction of being the one person I've written about in three consecutive books, with a different groundbreaking idea in each. Like the best ideas in science, this hypothesis has a simplicity that cuts straight through layers of complexity to the quick. It may not be right, for not all the great ideas in science are. But even if it's wrong, it shows how things *could* have come to be the way they are, and by suggesting experiments to test it, guides researchers in the right direction. It offers both insight and stimulus.

Many bacteria switch genes on and off in response to changes in their environment, says Allen; this in itself is common lore. One of the most important environmental switches is the presence or absence of raw materials. By and large, bacteria don't waste energy building new proteins to process raw materials if there aren't any around; they just close down the works until further notice. So Allen pictures a fluctuating environment – perhaps a stromatolite in shallow seas, in the vicinity of a hydrothermal vent that issues hydrogen sulphide into the world. The conditions would vary according to the tides, currents, time of year, hydrothermal activity, and so on. The critical factor is that Allen's hypothetical bacteria should possess both of the two photosystems, as cyanobacteria do today, but unlike cyanobacteria only ever use one of them at once. When hydrogen sulphide is present, the bacteria switch on Photosystem I and use it to produce organic matter from carbon dioxide. They can incorporate this new matter to grow, reproduce, and so on. But when conditions change, and the stromatolites are left without raw materials, these bacteria switch over to Photosystem II. Now they give up making new organic matter (they don't grow or reproduce any more), but they can maintain themselves by using sunlight to make ATP directly until better times. Each photosystem has its own benefit, and each evolved via a series of simple steps, as we've seen.

But what happens if a hydrothermal vent dies, or shifting currents lead to protracted changes in the environment? The bacteria must now rely on the electron circuit of Photosystem II for most of the time. But here there is a potential problem: the circuit can bung up with electrons from the environment, even if it only happens slowly in the electron-poor surroundings. The electron circuit is a bit like a pass-the-parcel game. An electron carrier either has an electron or it doesn't, just as a child either has a parcel when the music stops, or does not. But now imagine a rogue supervisor with a pile of parcels; he keeps passing them into the children's circle one at a time. In the end all the children end up with a parcel each. Nobody can pass on their parcel; the game grinds to a halt amid confusion.

Much the same happens with Photosystem II. The problem is inherent in sunlight, especially in the days before an ozone layer, when more ultraviolet radiation penetrated down to sea level. Ultraviolet rays not only split water, but can also throw off electrons from metals and minerals dissolved in the oceans, first among them manganese and iron. And that would have presented exactly the kind of problem that stymied our pass-the-parcel game. A trickle of electrons enters the circuit.

Neither iron nor manganese is found in high concentration in seawater today, for the oceans are thoroughly oxidised; but in ancient times, both were plentiful. Manganese, for example, is found in massive amounts on the sea floor in the form of curious cone-shaped 'nodules', which accrete over millions of years around objects like sharks' teeth, one of the few bits of living things that can withstand the intense pressure at the bottom of the ocean. There are thought to be a trillion tonnes of manganese-rich nodules scattered across the sea floor, a huge but uneconomic reserve. Even the more economic reserves, like the massive Kalahari manganese fields in South Africa (another 13.5 billion tonnes of ore), were precipitated from the oceans, 2,400 million years ago. In short, the oceans were once full of manganese.

For bacteria, manganese is a valuable commodity: it works as an antioxidant that protects cells against the destructive power of ultraviolet radiation. When a manganese atom absorbs a photon of ultraviolet radiation, it throws off an electron, becoming photo-oxidised, and in the process 'neutralises' the ray. The manganese is 'sacrificed' instead of more important bits and pieces of the cell such as proteins and DNA, which otherwise would be shredded by rays; and so bacteria welcome manganese into their abode with open arms. The trouble is that when these manganese atoms throw off an electron, it is always likely to be guzzled up by the 'grasping miser' form of chlorophyll in Photosystem II. And so the circuit gradually clogs up with electrons, just as our circle of children becomes swamped with parcels. Unless there is some way of bleeding off the excess electrons gumming up the circuit, Photosystem II becomes steadily less efficient.

How could bacteria bleed off electrons from Photosystem II? Here is the full genius of Allen's hypothesis. Photosystem II is clogged with electrons, while Photosystem I lies idle for lack of electrons. All that the bacteria need to do is disable the switch that prevents both photosystems from being active at once, either physiologically or by way of a single mutation. What happens then? Electrons enter Photosystem II from oxidised manganese atoms. They're then blasted up to a high energy level as the 'grasping miser' form of chlorophyll absorbs a ray of light. From here they're passed down the electron transport chain, using the release of energy to generate a little ATP. And then a diversion. Instead of returning to a gummed up Photosystem II, they are scavenged by the active Photosystem I, thirsty as it is for new electrons. Now the electrons are blasted up to a high energy level again, as the 'street hustler' form of chlorophyll absorbs a ray of light. And, of course, from here the electrons are finally passed to carbon dioxide, to generate new organic matter.

Does all this sound familiar? I have just described the Z scheme again. Just one single mutation connects the two photosystems in series, with electrons passing from manganese atoms, via the full Z scheme, to carbon dioxide to make sugars. What only now appeared to be an enormously convoluted and elaborate process is suddenly rendered virtually inevitable by a single mutation. The logic is flawless, the molecular parts are all in place and all serve a purpose as individual units. The environmental pressures are reasonable and predictable. Never did a single mutation make a bigger difference to the world!

It's worth a quick recap to appreciate the big picture in full relief. In the beginning there was a single photosystem, which probably used sunlight to extract electrons from hydrogen sulphide, and thrust them on to carbon dioxide to form sugars. At some point the gene became duplicated, perhaps in an ancestor of cyanobacteria. The two photosystems diverged under different usage.[9] Photosystem I carried on doing exactly what it had done before, while Photosystem II became specialised to generate ATP from sunlight by way of an electron circuit. The two photosystems were switched on and off according to the environment, but the pair were never switched on at the same time. Over time, however, Photosystem II has a problem, resulting from the properties of a circuit of electrons — any extra input of electrons from the environment jams up the circuit. It's likely there was a constant slow input of electrons from manganese atoms, used by bacteria to protect against ultraviolet radiation. One solution was to inactivate the switch, enabling both photosystems at once. Electrons would then flow from manganese, through both photosystems, to carbon dioxide, via a complex pathway that foreshadows the convoluted Z scheme in every eccentric detail.

We're now only one step short of full-blown oxygenic photosynthesis. We're drawing electrons from manganese, not from water. So how did the final shift occur? The surprising answer is that virtually nothing needed to change.

The oxygen-evolving complex is the nutcracker that pinions water 'just so' to have its electrons cracked out one by one. When all the electrons are removed, the invaluable waste, oxygen, is flushed out into the world. The oxygen-evolving complex is really a component of Photosystem II, but sits at the very edge, facing the outside world, and gives a sense of being 'tacked on'. It's shockingly small. The complex is a cluster of four manganese atoms and a single calcium atom, all held together by a lattice of oxygen atoms. And that's that.

For some years, the irrepressible Mike Russell, whom we already met in Chapters 1 and 2, has argued that the structure of this complex is remarkably similar to some minerals cooked up in hydrothermal vents, such as hollandite or tunnel calcium manganite. But until 2006, we didn't know the structure of

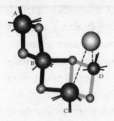

Figure 3.4 The ancient mineral structure of the oxygen-evolving complex – four manganese atoms (labelled A–D) linked by oxygen in a lattice, with a calcium atom nearby, as revealed by X-ray crystallography.

the manganese cluster in atomic resolution, and Russell's was a voice in the wilderness. But now we know. And although Russell was not quite right, his broad conception was absolutely correct. The structure, as revealed by a team headed by Vittal Yachandra at Berkeley, bears a striking resemblance to the mineral forms proposed by Russell (see Fig. 3.4).

Whether the original oxygen-evolving complex was simply a bit of mineral that got wedged in Photosystem II, we don't know. Perhaps manganese atoms became bound to oxygen in a lattice as they were oxidised by ultraviolet radiation, seeding the growth of a tiny crystal on site.[10] Perhaps the proximity of this cluster to chlorophyll, or to adjacent bits of protein, distorted it a little in some way, optimising its function. But whatever the origin of the cluster, there is a huge sense of the accidental about it. It is far too close to a mineral structure to be the product of biology. Like a few other metal clusters found at the heart of enzymes, it is almost certainly a throwback to the conditions found billions of years ago in a hydrothermal vent. Most precious of all jewels, the metal cluster was wrapped in a protein and held in trust for all eternity by the cyanobacteria.

However it formed, this little cluster of manganese atoms opened up a new world, not only for the bacteria that first trapped it, but for all life on our planet. Once it formed, this little cluster of atoms started to split water, the four oxidised manganese atoms combining their natural avidity to yank electrons from water, thereby releasing oxygen as waste. Stimulated by the steady oxidation of manganese by ultraviolet radiation, the splitting of water would have been slow at first. But as soon as the cluster became coupled to chlorophyll, electrons would have started to flow. Getting faster as chlorophyll became adapted to its task, water was sucked in, split open, its electrons drawn out, oxygen discarded. Once a trickle, ultimately a flood, this life-giving flow of electrons from water is behind all the exuberance of life on earth. We must thank it twice – once for being the ultimate source of all our food, and then again for all the oxygen we need to burn up that food to stay alive.

It's also the key to the world's energy crisis. We have no need for two photosystems, for we're not interested in making organic matter. We only need the two products released from water: oxygen and hydrogen. Reacting them together again generates all the energy we'll ever need, and the only waste is water. In other words, with this little manganese cluster, we can use the sun's energy to split water, and then react the products back together again to regenerate water – the hydrogen economy. No more pollution, no more fossil fuels, no more carbon footprints, no more anthropogenic global warming, albeit still some danger of explosions. If this little cluster of atoms changed the makeup of the world long ago, knowing its structure should be the first step to changing our own world today. As I write, chemists around the world are racing to synthesise this tiny manganese cluster in the lab, or something similar that works as well. Soon, surely, they will succeed. And then it can't be long before we learn to live on water and a splash of sunshine.

4

THE COMPLEX CELL

A Fateful Encounter

'The botanist is he who can affix similar names to similar vegetables, and different names to different ones, so as to be intelligible to every one,' observed the great Swedish taxonomist Carolus Linnaeus, a botanist himself. It might strike us as a limited ambition today, but in classifying the living world according to the traits of species, Linnaeus laid the foundations of modern biology. He was certainly proud of his achievements. 'God creates, Linnaeus organises,' he liked to say; and he would doubtless think it only proper that scientists still use his system today, subdividing all life into kingdom, phylum, class, genus and species.

This urge to categorise, to draw order from chaos, begins to make sense of the world around us and lies at the root of a number of sciences. Where would chemistry be without its periodic table? Or geology, without its eras and epochs? But there is a striking difference with biology. Only in biology is such classification still an active part of mainstream research. How the 'tree of life', that great chart plotting out the relatedness of all living organisms, should be structured is the source of rancour, even rage, among otherwise mild-mannered scientists. One article by Ford Doolittle, most urbane of scientists, conveys the mood in its title – 'Taking an axe to the tree of life'.

The problem is not one of arcane subtleties, but concerns the most important of all distinctions. Like Linnaeus, most of us still instinctively divide the world into plants, animals and minerals – they are, after all, the things we can

88

see. And what could be more different? Animals charge around, guided by their sophisticated nervous systems, eating plants and other animals. Plants produce their own matter from carbon dioxide and water, using the energy of sunlight, and are rooted to the spot; they have no need of a brain. And minerals are plainly inanimate, even if the growth of crystals persuaded Linnaeus, a touch embarrassingly, to categorise them too while he was at it.

The roots of biology as a subject likewise split into zoology and botany, and for generations never the twain did meet. Even the discovery of microscopic life forms did little to break down the old division. 'Animalcules' like amoeba, which move around, were dropped into the animal kingdom, and later on took the name *protozoa* (literally, 'first animals'), while coloured algae and bacteria were added to the plants. Yet if Linnaeus were to be pleased to find his system still in use, he would be shocked by the degree to which he had been deceived by outward appearances. Today the gap between plants and animals is perceived as quite narrow, while a dreadful gulf has opened up between bacteria and all the rest of complex life. It is the crossing of this gulf that causes so much disagreement among scientists: how exactly did life go from the primitive simplicity of bacteria to the complexity of plants and animals? Was it always likely to happen, or shatteringly improbable? Would it happen elsewhere in the universe, or are we more or less alone?

Lest the uncertainty play into the hands of those who would like to 'add a little God' to help out, there is no shortage of plausible ideas; the problem lies in the evidence, and specifically in the interpretation of evidence relating to deep time, a time perhaps 2,000 million years ago, when the first complex cells are thought to have emerged. The deepest question of all relates to why complex life arose only once in the whole history of life on our planet. All plants and animals are undoubtedly related, meaning that we all share a common ancestor. Complex life did not emerge repeatedly from bacteria at separate times – plants from one type of bacteria, animals from another, fungi or algae from yet others. On the contrary, on just one occasion a complex cell arose from bacteria, and the progeny of this cell went on to found all the great kingdoms of complex life: the plants, animals, fungi and algae. And that progenitor cell, the ancestor of all complex life, is very different from a bacterium. If we think about the tree of life, it's as if bacteria make up the roots, while the familiar complex organisms make up the branches. But whatever happened to the trunk? While we might consider single-celled protists, such as amoeba, to be intermediate forms, they are in fact in many respects nearly as complex as plants and animals. Certainly they sit on a lower branch, but they're still well above the trunk.

The gulf between bacteria and everything else is a matter of organisation at the level of cells. In terms of their morphology at least – their shape, size and contents – bacteria are simple. Their shape is usually plain, spheres or rods being most common. This shape is supported by a rigid cell wall around the outside of the cell. Inside there is little else to see, even with the power of an electron microscope. Bacteria are pared down to a minimum compatible with a free-living lifestyle. They are ruthlessly streamlined, everything geared for fast replication. Many keep as few genes as they can get away with; they have a propensity to pick up extra genes from other bacteria when stressed, bolstering their genetic resources, and then lose them again at the first opportunity. Small genomes are copied swiftly. Some bacteria can replicate every 20 minutes, enabling exponential growth at astounding rates, so long as raw materials last. Given sufficient resources (obviously an impossible demand) a single bacterium weighing a trillionth of a gram could found a population with a weight equal to that of the earth itself in less than two days.

Now consider complex cells, which rejoice in the formidable title *eukaryotes*. I wish they had a friendlier name, for their importance is second to none. Everything that is anything on this earth is eukaryotic – all the complex life forms that we've been talking about. The name derives from the Greek, *eu* meaning 'true' and *karyon* meaning 'nut', or 'nucleus'. Eukaryotic cells, then, have a true nucleus, distinguishing them from bacteria, which are termed *prokaryotes* for the lack of one. In a sense, the prefix pro- is a value judgement, for it proclaims the prokaryotes evolved before the eukaryotes. I think this is almost certainly true, but a few researchers would disagree. Regardless of exactly when it evolved, though, the nucleus is the defining feature of all eukaryotic cells. We can't hope to explain their evolution without understanding how and why the nucleus came to be, and conversely why no bacteria ever developed a true nucleus.

The nucleus is the 'command centre' of the cell, and is packed with DNA,

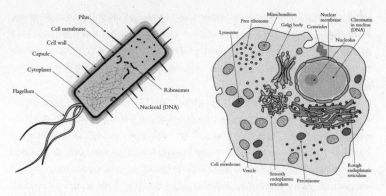

Figure 4.1 Differences between prokaryotic cells like bacteria, and complex eukaryotic cells with 'things inside' including a nucleus, organelles and internal membrane systems. This is emphatically *not* drawn to scale – eukaryotes are on average 10,000 to 100,000 times the volume of bacteria.

the stuff of genes. Beyond its very existence, there are several aspects of the eukaryotic nucleus that are alien to bacteria. Eukaryotes don't have a single circular chromosome, like bacteria, but a number of straight chromosomes, often doubled in pairs. The genes themselves are not strung out along the chromosome like beads on a string, as in bacteria, but are broken up into bits and pieces, with great expanses of non-coding DNA in between: we eukaryotes have 'genes in pieces', for whatever reason. And finally, our genes don't lie 'naked', like those of bacteria, but are fantastically bound in proteins, an arrangement as impervious to tampering as modern plastic gift-wrapping.

Outside the nucleus, too, eukaryotic cells are a world apart (see Fig. 4.1). They are usually much bigger than bacteria – on average, 10,000 to 100,000 times their volume. And then they are full of all sort of things: stacks of membranes; sealed vesicles galore; and a dynamic internal cell skeleton, which provides structural support, while at once being able to dismantle and rebuild itself around the cell, enabling changes of shape and movement. Perhaps most important of all are the organelles. These microscopic organs are devoted to particular tasks in the cell, just as the kidney or liver carry out their own specialised tasks in the human body. Most significant are the mitochondria, known as the 'powerhouses' of the cell, which generate energy in the form of ATP. An average eukaryotic cell harbours a few hundred mitochondria, but some contain as many as 100,000. Once upon a time they were free-living bacteria, and the consequences of their entrapment will loom large in this chapter.

These are merely differences in appearance. In behaviour, eukaryotic cells are equally arresting, and again utterly different from bacteria. With a few fiddling exceptions, so to speak, practically all eukaryotes have sex: they generate sex cells like the sperm and egg, which fuse together to form a hybrid cell with half the genes of the father, and half of the mother (more on this in the next chapter). All eukaryotic cells divide via a spellbinding gavotte of chromosomes, which double up and align themselves on a spindle of microtubules, before retiring to opposite ends of the cell, as if with a bow and a curtsy. The list of eukaryotic eccentricities goes on, and I want to mention just one more: phagocytosis, or the ability to gobble up whole cells and digest them within. This trait seems to be an ancient one, even if a few groups, such as fungi and plants, have lost it again. So, for example, although most animal and plant cells don't troop around engulfing other cells, immune cells do exactly that when they consume bacteria, drawing on the same apparatus as an amoeba.

All this relates equally to all eukaryotic cells, whether plant, animal or amoeba. There are, of course, many differences between them too, but set against their shared properties, these pale into insignificance. Many plant cells contain chloroplasts, for example, organelles responsible for photosynthesis. Like mitochondria, chloroplasts were once free-living bacteria (in this case cyanobacteria), which were swallowed whole by a common ancestor of all plants and algae. For whatever reason, this ancestral cell failed to digest its dinner, and through a case of indigestion acquired everything needed to become self-sufficient, powered only by sun, water and carbon dioxide. In one gulp, it set in motion the entire train of circumstances that ultimately separates the stationary world of plants from the dynamism of animals. Yet peer within a plant cell, and this is but a single difference set against a thousand traits in common. We could go on. Plants and fungi rebuilt outer cell walls to reinforce their structure; some have vacuoles, and so on. But these are all no more than trifling differences, as nothing compared to the empty void that separates eukaryotic cells from bacteria.

Yet it is a teasing void, at once real and imaginary. There is some degree of overlap between bacteria and eukaryotic cells in almost all of the traits we've considered. There are a few large bacteria, and a good many tiny eukaryotes: their size range overlaps. Bacteria have an internal cell skeleton, alongside their cell wall, composed of very similar fibres to the eukaryotic cell skeleton. It even appears to be dynamic, to a point. There are bacteria with straight (not circular) chromosomes, with structures that resemble a nucleus, with internal membranes. A few lack cell walls, at least for part of their life cycle. Some live in sophisticated colonies that might pass as multicellular organisms, certainly for bacterial apologists. There are even one or two cases of bacteria harbouring other, even smaller, bacteria inside them – an enigmatic finding, given that no bacterium is known that can swallow cells by phagocytosis. My sense is that bacteria made a start along the trail to almost all eukaryotic traits, but then stopped short, unable to continue the experiment, for whatever reason.

You may feel, not unreasonably, that an overlap is the same thing as a continuum, and therefore there is nothing to explain. There can be no void between bacteria and eukaryotes if there is a continuum from simple bacteria at one end of the spectrum to complex eukaryotes at the other. This is true in a sense, but I think it is misleading, for although there is indeed some degree of overlap, it is really an overlap of two separate spectra – a truncated one for bacteria, which runs from 'extreme simplicity' to 'limited complexity', and a vastly longer one for eukaryotes, which runs from 'limited complexity' to 'mind-boggling complexity'. Yes, there is overlap, but bacteria never made it very far up the eukaryotic continuum; only the eukaryotes did that.

The difference is illustrated forcibly by history. For the first 3,000 million years or so of life on earth (from 4,000 to 1,000 million years ago), bacteria dominated. They changed their world utterly, yet barely changed themselves. The environmental changes brought about by bacteria were awesome, on a scale that even we humans find hard to conceive. All the oxygen in the air, for example, derives from photosynthesis, and early on from cyanobacteria alone. The 'great oxidation event', when the air and sunlit surface oceans became flooded with oxygen, around 2,200 million years ago, transfigured our planet forever; but the shift didn't make much of an impression on bacteria. There was merely a shift in ecology, towards the kind of bacteria that like oxygen. One type of bacteria came to be favoured over another, but all of them remained resolutely bacterial. Exactly the same is true of other monumental shifts in conditions. Bacteria were responsible for suffocating the ocean depths with hydrogen sulphide, for a little matter of 2,000 million years; but they always remained bacteria. Bacteria were responsible for oxidising atmospheric methane, precipitating a global freeze, the first snowball earth; but they remained bacteria. Perhaps the most significant change of all was brought about by the rise of complex multicellular eukaryotes, in the last 600 million years. The eukaryotes offered up new ways of life for bacteria, like causing infectious diseases; but bacteria are still bacteria. Nothing is more conservative than a bacterium.

And so history started with the eukaryotes. For the first time, it became possible to suffer 'one damned thing after another', instead of the interminable sameness of it all. On a few occasions, things happened damned fast. The Cambrian explosion, for example, is an archetypal eukaryotic affair. This was the moment – a geological moment, lasting perhaps a couple of million years – when large animals abruptly materialised in the fossil record for the first time. These were not morphologically tentative forms, this no procession of worms, but an astonishing catwalk of weird body plans, some of which vanished again almost as swiftly as they had appeared. It was as if a deranged creator had woken up with a start and immediately set about making up for all those aeons of lost time.

The technical term for such an explosion is a 'radiation', in which one particular form suddenly takes off, for whatever reason, and embarks on a short period of unbridled evolution. Inventive new forms radiate out from the ancestral form like the spokes on a wheel. While the Cambrian explosion is the best known, there are many other examples: the colonisation of the land, the rise of flowering plants, the spread of grasses, or the diversification of mammals, to mention but a few. These events tend to occur when genetic promise comes face to face with environmental opportunity, as in the wake of a mass extinction. But regardless of the reason, such magnificent radiations are uniquely eukaryotic. Each time, only eukaryotic organisms flourished;

bacteria, as ever, remained bacteria. One is forced to conclude that human intelligence, consciousness, all the properties that we hold so dear and seek elsewhere in the universe, simply could not arise in bacteria: on earth, at least, they are uniquely eukaryotic traits.

The distinction is sobering. While the bacteria put us eukaryotes to shame with the cleverness of their biochemistry, they are seriously stunted in their morphological potential. They seem to be incapable of producing the marvels we see around us, whether hibiscus or hummingbird. And that makes the transition from simple bacteria to complex eukaryotes perhaps the single most important transition in our planet's history.

Darwinians don't like gaps much. The conception of natural selection as a gradual series of tiny steps, each an improvement over the last, implies that we should see far more intermediates than we actually do. Darwin himself tackled this perceived difficulty in *The Origin*, with the simple observation that all the interim steps are, by definition, less well adapted than the 'endpoints' that we see around us today. By the very nature of selection, worse forms will lose out to better rivals. Obviously a bird capable of proper flight will fare better than any relatives obliged to get by with 'stumps' instead of wings. In the same way, new computer software displaces obsolete versions from the market; when did you last see a Windows 286 or 386 operating system? They were the state of the art once, just as prototype wings must have been in their own context (and indeed are today for flying squirrels or gliding snakes). But with the lapse of time, these early operating systems disappeared without trace, leaving an apparent 'void' before, say, Windows XP.[1] We accept that Windows operating systems have improved over time, but if we look for evidence of this evolution simply by comparing computers in use today, we'd find little sign of it, except for a few dusty fossils tucked away in the attic. Likewise, with life, if we want to find evidence of a continuum we have to look to the fossil record, to the period when changes were taking place.

The fossil record is certainly patchy, but there are far more intermediate forms than a vociferous minority of zealots are willing to admit. When

Darwin was writing, there really was a 'missing link' between apes and human beings: no fossil hominids with intermediate features were known. But over the last half-century, palaeontologists have recovered scores of them. By and large, they all fall exactly where one would expect on a spectrum of traits, such as brain size or gait. Far from an absence of intermediate forms, we are now faced with an embarrassment of riches. The difficulty is that it's hard to know which of these hominids, if any, are the ancestors of modern human beings, and which just disappeared without issue. Because we don't know all the answers (yet) we still hear loud claims that the missing link has *never* been found, which is a violation of honesty and truth.

But from my point of view, as a biochemist, fossils are a beautiful distraction. Given the sheer improbability and unpredictability of fossilisation, along with a systematic bias against soft-bodied creatures such as jellyfish, and against plants and animals that live on dry land, fossils should *not* preserve an unblemished record of the past. If they did, we'd suspect skulduggery. When, occasionally, they do, we should celebrate it as a wondrous accident of fortune, a rare constellation of circumstances bordering on a miracle, yet ultimately no more than a pleasing corroboration of the real evidence for natural selection. That real evidence is all around us, in this age of genomics, in the sequences of genes.

Gene sequences preserve something much closer to the fabric of evolution than fossils ever can. Take any gene you want. Its sequence is a long parade of letters, the order of which encodes the succession of amino acids in a protein. There are typically a couple of hundred amino acids in a protein, each one of which is encoded by a triplet of letters in DNA (see Chapter 2). As we have already noted, eukaryotic genes often include long interludes of non-coding DNA, interspersed between shorter coding stretches. Adding it all up, the sequence of a gene is typically thousands of letters long. Then there are tens of thousands of genes, each set up in a similar way. All in all, a genome is a ribbon with millions or billions of letters, the order of which has a great deal to say about the evolutionary heritage of its owner.

The same genes, encoding proteins responsible for the same tasks, are found in diverse species, from bacteria to mankind. Over evolutionary time, detrimental mutations in gene sequences are weeded out by selection. This has the effect of retaining the same letters at equivalent positions in a gene sequence. From a purely practical point of view, that in turn means we can recognise related genes in different species despite the passage of unimaginable aeons. As a rule of thumb, though, only a small proportion of the thousands of letters in a gene are particularly important; the rest can vary more or less freely as mutations build up over time, because the changes don't matter much and so are not eliminated by selection. The more time that passes, the more these mutations accumulate, and so the more distinct two gene sequences become. Species that share a relatively recent common ancestor, such as chimps and humans, have numerous gene sequences in common, then, while those with a more distant common ancestor, such as daffodils and humans, retain less. The principle is much the same with languages, which drift apart over time, steadily losing any semblance of common ancestry but for a few points of hidden similarity that still tie them together.

Gene trees are based on the differences in gene sequence between species. Although there is a degree of randomness about the accumulation of mutations, these balance out if averaged over thousands of letters, giving a statistical probability of relatedness. Using a single gene, we can reconstruct the family tree of all eukaryotic organisms with a degree of precision that is beyond the wildest dreams of fossil hunters. If you harbour any doubts, simply repeat the analysis with a different gene, and see if you reproduce the same pattern. Because eukaryotic organisms have hundreds, if not thousands, of genes in common, the approach can be repeated again and again, and the single trees superimposed over each other. With a little computing power, a 'consensus' tree can be built, giving the most probable relationship between all eukaryotic organisms. Such an approach is a far cry from gaps in the fossil record: we can see exactly how we are related to plants, fungi, algae, and so on (see Fig. 4.2). Darwin knew nothing about genes, but it is the fine structure of genes, more than anything else, that has eliminated all the distasteful gaps from the Darwinian view of the world.

So far so good, but there are some problems too. These are largely caused by statistical errors in the measurement of change over deep time. The basic trouble is that there are only four different letters in DNA, and mutations (at least of the type we're interested in here) normally replace one letter with a different letter. If most letters are only replaced once, that's fine; but over the

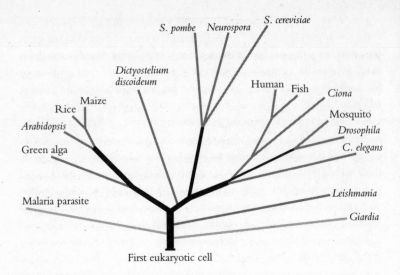

Figure 4.2 Conventional tree of life, showing the divergence of eukaryotic organisms from a common ancestor, a single-celled organism living perhaps around 2 billion years ago. The longer the branch, the greater the evolutionary distance, which is to say the more different are the genes.

vast tracts of evolutionary time, many will be replaced more than once. As every change is a lottery, it's hard to know whether each letter has been replaced once or five or ten times. And if a letter hasn't changed at all, it might be that it was never replaced, or that it was replaced several times, each time with a 25 per cent chance of restoring the original letter. Because such analyses are a matter of statistical probability, there comes a point when we can't discriminate between the alternative possibilities. And as bad luck would have it, the point at which we flounder in a sea of statistical doubt corresponds roughly to the emergence of the eukaryotic cell itself. The crucial transition from bacteria to eukaryotes is submerged in a tide of genetic uncertainty. The only way out of the problem is to use a more finely grained statistical sieve – to select our genes with more care.

✳

In eukaryotic cells there are two broad classes of gene: those with equivalents in bacteria and those that seem to be uniquely eukaryotic. Nothing like them has yet been found in the bacterial world.[2] These unique genes are known as 'eukaryotic signature genes', and their heritage is the source of bitter dispute. Some say that they prove the eukaryotic domain is as venerable as the bacteria. The fact that so many eukaryotic genes are distinctive, they say, surely means that the eukaryotes have been distancing themselves from bacteria since the beginnings of time. If their rate of divergence is assumed to be constant (the steady 'tick, tick' of the mutation rate acting as a molecular clock), the scale of differences would have us believe that the eukaryotes evolved more than 5 billion years ago, at least half a billion years before the earth even formed at all. Shurely shome mishtake, as the English satirical magazine *Private Eye* would say.

Others say that the eukaryotic signature genes tell us nothing about the evolutionary heritage of the eukaryotes, as we have no way of knowing how rapidly genes might have been evolving in the distant past, and no reason to assume that there is anything clocklike about their divergence. Certainly we know that some genes evolve faster than others today. And the fact that molecular clocks point to such a dubious antiquity suggests that either life was seeded from space – a cop-out in my opinion – or the clocks are wrong. Why would they be so wrong? Because the rate at which genes evolve depends a lot on the circumstances, and especially on the kind of organism that they find themselves in. As we've seen, bacteria are the ultimate conservatives, remaining forever bacteria, whereas eukaryotes appear prone to episodes of spectacular change, as in the Cambrian explosion. Arguably no episode was more dramatic, from the genes' point of view, than the formation of the eukaryotic cell itself, and if so there is every reason to expect a furious rate of change in those early formative days. If eukaryotes evolved more recently than bacteria, as most researchers believe, then their genes are very different because, for a time, they evolved very fast, mutating, recombining, duplicating and mutating again.

The eukaryotic signature genes have little to tell us, then, about the evolution of the eukaryotes. They've simply evolved so fast and so far that their origins have been lost in the fog of time. So what of the second group of genes, those that do have known equivalents in bacteria? These are immediately more informative, because we can begin to compare like with like. Genes that are found in both bacteria and eukaryotes often encode core processes in the cell, whether core metabolism (the way in which energy is generated and used to build the key building blocks of life, such as amino acids and lipids) or core informational processes (the way that DNA is read off and translated into the active currency of proteins). Such core processes usually evolve slowly, because much else depends on them. Change a single aspect of protein synthesis, and you alter the manufacture of all proteins, not just one of them. Likewise, change energy generation even slightly, and you may jeopardise the whole operation of the cell. Because changes in core genes are more likely to be punished by selection, these genes evolve slowly, and so should give us a finer-grained analysis of evolution. A tree built from such genes could, in principle, illuminate how eukaryotes relate to bacteria. It might indicate which group they arose from, and perhaps even hint as to why.

The American microbiologist Carl Woese first built such a tree in the late 1970s. He chose a gene encoding part of the core informational processes of the cell – specifically, part of the tiny molecular machines, called ribosomes, that carry out protein synthesis. For technical reasons, Woese didn't originally use the gene itself, but rather an RNA copy, which is read off from the gene and incorporated directly into the ribosome. He isolated this ribosomal RNA from various bacteria and eukaryotes, sequenced it, and built a tree by comparing the sequences. The findings were a shock, challenging long-held ideas about how the living world should be structured.

Woese found that all life on our planet falls into three big groups, or domains (see Fig. 4.3). The first group is the bacteria, as we might expect, and the second is the eukaryotes. But a third group, now known as the archaea, came from nowhere to assume prominence on the world stage. Although a handful of archaea had been known about for a century, until Woese's new tree they were just seen as a small sect within the bacteria. After Woese, they became as important as the eukaryotes, despite the fact that they *look* exactly like bacteria: they are tiny, usually have an external cell wall, lack a nucleus,

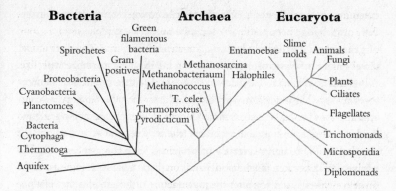

Figure 4.3 Tree of life based on ribosomal RNA, showing Carl Woese's split into the three great domains of life: the bacteria, the archaea and the eukaryotes.

or anything at all worthy of comment inside, and never group into colonies that might be mistaken for multicellular organisms. Inflating their importance struck many as an impudent restructuring of the world, relegating all the marvellous diversity of plants, animals, fungi, algae and protists to an insignificant corner of a tree dominated by prokaryotic cells. All the manifold differences between plants and animals, Woese challenged us to believe, are as nothing compared to the invisible chasm between bacteria and archaea. Biologists of the stature of Ernst Mayr and Lynn Margulis were outraged. Reflecting on the sharp exchanges years later, the journal *Science* hailed Woese as 'microbiology's scarred revolutionary'.

Yet for all that, most researchers do now accept Woese's tree, or at least the prominence of the archaea. At a biochemical level they really are different from bacteria, in almost every respect. The cell membrane is composed of dissimilar lipids, synthesised with the aid of a different set of enzymes. Their cell wall has nothing in common with the bacterial cell wall. Their metabolic pathways have little overlap with bacteria. As we saw in Chapter 2, the genes controlling DNA replication are quite unrelated. And now that analyses of whole genomes are commonplace, we know that the archaea share less than a third of their genes with bacteria; the rest are unique. All in all, Woese's unanticipated RNA tree served to highlight a series of major biochemical differences between bacteria and archaea, albeit discreet to the point of invisibility, which together uphold his bold reclassification of life.

The second unexpected aspect of Woese's tree was the surprisingly close relationship between archaea and eukaryotes: both share a common ancestor only distantly related to bacteria (see Fig 4.3). In other words, the common ancestor of archaea and eukaryotes branched away from bacteria very early in evolution, and later on split to form the modern archaea and eukaryotes. Again, Woese's tree has been upheld by biochemistry, at least in several important respects. In their core informational processes, in particular, archaea and eukaryotes have much in common. Both wrap up their DNA in remarkably similar proteins (histones), both replicate and read off their genes in a comparable way, and both build proteins by a common mechanism, all of which differ in detail from bacteria. In these respects the archaea amount to a missing link, in part straddling the gulf between bacteria and eukaryotes. In essence, the archaea are allied to bacteria in appearance and behaviour, but have some startlingly eukaryotic traits in the way they handle DNA and proteins.

The trouble with Woese's tree is that it is built from a single gene, and so loses the statistical power gained by superimposing gene trees. We can only rely on a single-gene tree if we can be certain that the gene chosen reflects the true heritage of eukaryotic cells. The best way to test whether that is indeed the case is to superimpose other slowly evolving genes, to see whether they, too, replicate the same deep branching pattern. When this is done, the answers become confounding. If we choose only genes shared by all three domains of life (those found in bacteria, archaea and eukaryotes), we can reconstruct robust trees for bacteria and archaea, but not for eukaryotes. The eukaryotes are a confusing mix. Some of our genes apparently derive from archaea, others from bacteria. The more genes that we study – and one recent analysis combined 5,700 genes, drawn from 165 different species into a 'supertree' – the more plain it becomes that the eukaryotic cell did not evolve in a standard 'Darwinian' way, but rather by some sort of mammoth gene fusion. From a genetic point of view, the first eukaryote was a chimera – half archaea, half bacteria.

*

According to Darwin, life evolves by a slow accumulation of differences over time, as different lineages diverge away from their common ancestor. The outcome is a branching tree, and there is no doubt that such trees are the best way of depicting the evolution of most of the organisms we can see, essentially, most large eukaryotes. But it is equally clear that a tree is not the best way to depict the evolution of microbes, whether bacteria, archaea or eukaryotes.

There are two processes that confound Darwinian gene trees: lateral gene transfers and whole genome fusions. For microbial phylogenists, trying to make sense of the relationships between bacteria and archaea, lateral gene transfers are distressingly common. This rather cumbersome term simply means that genes are passed around, like cash, from one organism to another. The outcome is that the genome a bacterium passes on to its daughter may or may not be similar to the genome it inherited from its own parent. Some genes, like Woese's ribosomal RNA, tend to be passed down 'vertically' from one generation to the next, while others are swapped around, often to-and-from quite unrelated microbes.[3] The overall picture is somewhere between a tree and a network, with core genes (like ribosomal RNA) tending to build a tree, others tending to form a net. Whether there exists a core group of genes that is *never* shuffled around by lateral gene transfer is a bone of contention. If not, the very idea of tracing eukaryotic evolution back to particular groups of prokaryotes becomes a nonsense. Such 'groups' only have a historical identity if they inherit traits chiefly from their own ancestors, rather than other random groups. Conversely, if a small core of genes is *never* passed around but all other genes *are*, what does that say about identity? Is *E. coli* still *E. coli* if 99 per cent of its genes are replaced at random?[4]

Genome fusions raise similar difficulties. Here the problem is that the Darwinian tree is turned on its head: instead of divergence, we have convergence. The question then becomes which of the two (or more) partners reflects the true course of evolution? If we track only the gene for ribosomal RNA, we recover a conventionally branching Darwinian tree; but if we consider a large number of genes, or whole genomes, we recover a ring, in which the branches that had previously diverged now converge again and fuse together (see Fig. 4.4).

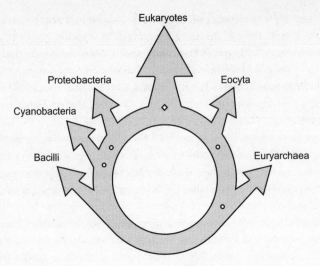

Figure 4.4 The 'ring of life'. The last common ancestor of all life is at the bottom, and splits into bacteria (left) and archaea (right), representatives of which fuse again to give rise to the chimeric eukaryotes at the top.

There is no doubt that the eukaryotic cell is a genetic chimera; nobody questions that evidence. The question that divides camps now is just how much emphasis should be placed on standard Darwinian evolution, and how much on drastic gene fusion? Put another way, how many of the properties of the eukaryotic cell evolved by gradual evolution of a host cell, and how many of them could *only* develop *after* a gene fusion? Over many decades, scores of theories have been advanced on the origin of the eukaryotic cell, ranging from pure imagination, if not fabrication, to carefully reconstructed biochemistry. None has yet been proved. All fall into two big groups, emphasising either slow Darwinian divergence or dramatic gene fusion. The groups correspond to two sides of an older struggle in biology, between those who argue that evolution proceeds by gradual continuous change, and those who insist that immense periods of stasis, or equilibrium, are punctuated occasionally by sudden and dramatic change. As the old jibe has it, evolution by creeps versus evolution by jerks.[5]

In terms of the eukaryotic cell, Christian de Duve dubs the two groups the 'primitive phagocyte' and the 'fateful encounter' hypotheses. The primitive phagocyte idea is conceptually Darwinian, and is championed most persuasively by the Oxford evolutionist Tom Cavalier-Smith and by de Duve himself. The essential idea is that the ancestor of the eukaryotic cell gradually accrued all the characteristics of the modern eukaryotic cell, including a nucleus, sex, a cell skeleton and, most importantly of all, phagocytosis, or the ability to go around swallowing up other cells by changing shape, engulfing them and digesting inwardly. The only trait this primitive phagocyte lacked, relative to modern eukaryotes, was mitochondria, which generate energy using oxygen. Presumably it relied on fermentation for its energy, a far less efficient process.

But for a phagocyte, swallowing the ancestors of the mitochondria was just part of a good days' work. What could be easier? Indeed, how else could one cell get inside another? Certainly, the possession of mitochondria gave the primitive phagocyte an important advantage – they would have revolutionised its energy generation – but they didn't fundamentally change its makeup. It was a phagocyte before it swallowed mitochondria and remained a phagocyte afterwards, albeit one with more energy. Many genes from the enslaved mitochondria, however, would have been transferred over to the nucleus and incorporated into the host cell genome, and it is this transfer that accounts for the chimeric nature of the modern eukaryotic cell. The genes derived from the mitochondria are bacterial in heritage. So, the defenders of a primitive phagocyte don't dispute the chimeric nature of modern eukaryotes, but posit a non-chimeric phagocyte – a bona fide if primitive eukaryote – as the host cell.

Back in the early 1980s, Tom Cavalier-Smith highlighted a group of a thousand or more species of primitive-looking single-celled eukaryotes that lack mitochondria. Perhaps, he said, a few of them had survived from the early days of the eukaryotic cell, direct descendants of the primitive phagocyte that had never possessed mitochondria. If so, then they should not show any signs of genetic chimerism, as they would have evolved by purely Darwinian processes. But over the two ensuing decades, all did turn out to be chimeras: all, it seems, once possessed mitochondria, and later lost them, or modified them into something else. *All* known eukaryotic cells either possess mitochondria today, or once did in the past. If ever there existed a primitive phagocyte, lacking mitochondria, it didn't leave any direct descendants. That doesn't mean to say it never existed, merely that its existence is conjectural.

The second group of theories falls under the 'fateful encounter' flag. All of them assume that there was some kind of an association between two or more prokaryotic cells, which ended up forming a closely-knit commune of cells – a chimera. If the host cell was not a phagocyte, but an archaeon with a cell wall, the biggest question is how on earth did the other cells get inside? The main protagonists of this idea, notably Lynn Margulis and Bill Martin (whom we already met in Chapter 1) point to various possibilities. Margulis, for example, suggests that a bacterial predator could have forced an entry into the innards of other bacteria (and there are examples). Martin, in contrast, posits a mutual metabolic relationship between cells, which he explores in detail, whereby each partner exchanges raw materials with the other.[6] In this case, it's hard to see how one prokaryote can physically get inside another one without the aid of phagocytosis, but Martin furnishes two examples where exactly this has happened in bacteria (see Fig. 4.5).

The fateful encounter theories are all essentially non-Darwinian, in that they don't posit small changes as the mode of evolution, but the relatively dramatic origin of a new entity altogether. Crucially, the assumption is that all eukaryotic traits evolved only *after* the fateful union. The collaborating cells themselves were strictly prokaryotes, lacking phagocytosis, sex, a dynamic cytoskeleton, a nucleus, and so on. These traits only developed after the union was cemented. The implication is that there was something about the union itself which transformed the arch-conservative, never-changing prokaryote into its antithesis, the ultimate speed junky, the ever-changing eukaryote.

How can we distinguish between these possibilities? We've seen that the eukaryotic signature genes can't. We have no way of knowing whether they evolved over 4 billion years or 2 billion years, whether they evolved before a fusion with mitochondria or afterwards. Even the slowly evolving genes with prokaryotic counterparts are unreliable: it depends on which ones we choose. For example, if we consider Woese's ribosomal RNA tree, the data are

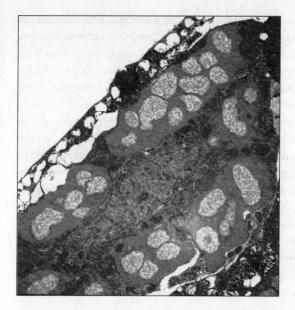

Figure 4.5 Bacterial cells living within other bacterial cells. Numerous gamma proteobacteria (pale mottled gray) living inside beta-proteobacteria (darker gray), all inside a eukaryotic cell, with the dappled nucleus in the bottom centre of the picture.

compatible with a primitive phagocyte model. The reason is that, in Woese's tree, eukaryotes and archaea are 'sister' groups that share a common ancestor; they have the same 'mother'. This means that eukaryotes did not evolve *from* archaea, any more than one sister begets another. The common ancestor in this case was almost certainly prokaryotic (otherwise all archaea must have lost their nucleus); but beyond that there's little we can say for sure. It's possible that the eukaryotic line evolved into a primitive phagocyte before engulfing the mitochondria, but there's not a shred of genetic evidence to support this conjecture.

Conversely, if we build more complex gene trees, using a larger number of genes, then the sister kinship between eukaryotes and archaea begins to break down; instead it looks as if the archaea actually begot the eukaryotes. Exactly which archaeon is uncertain, but the largest study so far – the one I've already mentioned that built a supertree from 5,700 genes – implies that the host cell was a true archaeon, perhaps most closely related to the modern thermoplasma. This difference is absolutely critical. If the host cell was a true archaeon (by definition, a prokaryote, lacking a nucleus, sex, a dynamic cell skeleton, phagocytosis, and so on), it was obviously not a primitive phago-cyte. And if that's the case, the 'fateful encounter' hypothesis must be true: the eukaryotic cell sprang from a union between prokaryotic cells. There never was a primitive phagocyte, and the absence of evidence for its existence somersaults into evidence for its absence.

Yet this is unlikely to be the final answer either. A great deal depends on precisely which genes or species are chosen, and the selection criteria. Each time these are varied, the tree reconfigures itself into a different branching pattern, confounded by statistical assumptions, lateral transfers among prokaryotes, or other unknown variables. Whether the situation is resolvable with more data or is simply unanswerable by genetics – a biological equiva-lent of the uncertainty principle in which the closer we get the fuzzier every-thing becomes – is frankly anybody's guess. But if the question can't be resolved with genetic data, are we condemned to perpetual mud-slinging between rival factions of scientists with clashing temperaments? There may be another way.

All known eukaryotic cells either have mitochondria now, or once had them in the past. And curiously, all mitochondria that still function as mitochondria, which is to say, that generate energy using oxygen, retain a handful of genes, a vestige of their former life as free-living bacteria. This tiny mitochondrial genome, I think, hides the deep secret of the eukaryotic cell.

The eukaryotes have been diverging for the best part of 2 billion years, and during that time have been independently losing their mitochondrial genes. All have lost between 96 and 99.9 per cent of their mitochondrial genes, prob-ably transferring the majority of them to the cell nucleus; but none has lost them all without at the same time losing the capacity to use oxygen. That doesn't sound random. Transferring all the mitochondrial genes to the nucleus is rational and tidy. Why retain hundreds of gene outposts in every cell, when 99.9 per cent of genes are stored as a single copy, along with a backup, in the nucleus? And retaining any genes at all in mitochondria means that the entire apparatus for reading them off and translating them into active proteins also has to be retained in every single mitochondrion. Such profligacy would bother accountants, and natural selection is, or ought to be, the patron saint of accountants.

The plot thickens. Mitochondria are a silly place to store genes. They are often glibly called the powerhouses of the cell, but the parallel is quite exact. Mitochondrial membranes generate an electric charge, operating across a few millionths of a millimetre, with the same voltage as a bolt of lightning, a thousand times more powerful than domestic wiring. To store genes here is like depositing the most precious books of the British Library in a dodgy nuclear power station. And the threat isn't just theoretical. Mitochondrial genes mutate far faster than genes in the nucleus. For example, in yeast, a handy experimental model, they mutate some 10,000 times faster. Yet despite this, it is critical that the two genomes (the nuclear and mitochondrial genomes) function properly together. The high-voltage force powering eukaryotic cells is generated by proteins encoded by both genomes. If they fail to function well together, the penalty is death – death for the cell, and death for the organism. So two genomes must work together to generate energy. Any failure to cooperate ends in death, but the mitochondrial genes are mutating at 10,000 times the rate of nuclear genes, making such tight collaboration close to impossible. This is surely the single most peculiar trait of eukaryotic cells. To dismiss it as a mere quirk, as textbooks tend to do, is to miss an Everest of a clue. If it were helpful to get rid of all mitochondrial genes, we can be certain that natural selection would have done so by now, at least in one species. Surely they're kept for a reason.

So why retain a mitochondrial genome then? According to the freethinking John Allen, whose ideas on photosynthesis we discussed in Chapter 3, the answer is simply to control respiration. No other reason is big enough. Respiration means different things to different people. For most people, it just means breathing. But for biochemists respiration refers to the minutiae of breathing at a cellular level, the series of tiny steps in which food is reacted with oxygen to generate an internal voltage with the force of lightning. I can't think of a selection pressure more immediate than breathing; and at the molecular level inside cells, it's the same. Cyanide, for example, blocks cell respiration and brings the workings of the cell to an end faster than a plastic bag over the head. Even when functioning normally, respiration has to be continuously fine-tuned by 'fiddling with the knobs', adjusting power to demand. Allen's critical point is that matching power to demand in this way requires constant feedback, which can only be achieved by the *local* control of gene activity. Just as an army's tactical disposition on the ground shouldn't be controlled by a remote central government, so the nucleus is not well placed to tune up or down the many hundreds of individual mitochondria in a cell. Mitochondria, then, retain a small genome to tune respiration, matching power to demand.

Allen's ideas are by no means proved, although evidence is building to support them. But if he is right, the connotations help explain the evolution of the eukaryotic cell. If a number of gene outposts are *needed* to control respiration in eukaryotic cells, then it stands to reason that a large complicated cell couldn't control respiration without them. Think now of the selection pressures facing bacteria and archaea. Both produce ATP in the same way as mitochondria, by generating an electrical charge over a membrane. Prokaryotes, however, use their outer cell membrane and that gives them a problem with size. Effectively they breathe over their skin. To understand why this is a problem, just think of peeling potatoes. If you need to peel a ton of potatoes, you should choose only the biggest ones; you get far more potato relative to skin that way. Conversely, you get far more skin with small potatoes. Bacteria are like potatoes that breath over the skin, and the bigger they get, the less they can breathe.[7]

In principle, bacteria could get around their breathing difficulties by internalising their energy-generating membranes, and in practice this happens to a degree, as we noted earlier on: some bacteria do have internal membranes, giving them a 'eukaryotic' look. But they never went very far: an 'average' eukaryotic cell has hundreds of times more internal membrane dedicated to energy generation than even the most energetic bacterium. As in so many other traits, bacteria started out up the eukaryotic spectrum, and then stopped short. Why? I suspect because they can't control respiration over a wider area of internal membranes. To do so they would need to hive off multiple sets of genes, as in the mitochondria, and that is not at all easy to achieve. All the selection pressures on bacteria – to replicate quickly, casting away all but a minimal genome – militate against large complex bacteria.

But that is exactly what phagocytosis demands. Phagocytes need to be large enough to engulf other cells; and to phagocytose, they need plenty of energy to move around, to physically change shape, and to engulf prey. The problem is that, as bacteria get larger, they become less energetic, less able to spare the energy for moving around and changing shape. It seems to me that a tiny bacterium, geared for fast replication, would prevail over an energetically challenged larger cell every time, well before the larger cell ever managed to evolve all the multifaceted attributes of a phagocyte.

But the 'fateful-encounter' hypothesis is a different matter. Here, two prokaryotes live alongside each other in mutual metabolic harmony, each providing the other with some service. Symbiotic relationships of this nature are common among prokaryotes, to the point of being the rule rather than the exception. What is extremely rare, but still a documented occurrence in prokaryotes, is the physical engulfment of one partner by the other. Once inside, the whole cell, including its internalised bacteria, now evolves as a single entity. While each continues to provide a service to the other, any supernumerary traits are gradually eroded away until the internalised bacteria are left doing little more than a few jobs for the host cell – energy generation, in the case of the bacteria that became mitochondria.

The immense advantage offered by the mitochondria, and the reason that mitochondria enabled the eukaryotic cell to evolve at all, is that they provided a ready-made system of internal energetic membranes, along with the outposts of genes needed to control respiration locally. Only when the host cell possessed mitochondria could it scale itself up to become a large active phagocyte without incurring a crippling energetic cost. If all that is true, then a primitive phagocyte lacking mitochondria never existed because phagocytosis is simply not possible without mitochondria.[8] The eukaryotic cell was forged in a union between two prokaryotic cells. The union relieved the energy constraints that forced bacteria to remain bacteria for all time. Once the constraints were eased, a new way of life – phagocytosis – became possible for the first time. The eukaryotic cell only evolved once because the union of two prokaryotes, in which one gains entry to another, is truly a rare event, a genuinely fateful encounter. All that we hold dear in this life, all the marvels of our world, stem from a single event that embodied both chance and necessity.

Early in this chapter, I noted that we can only understand or explain the origin of the eukaryotic cell when we have grasped the significance of its defining attribute, the nucleus itself; and so it is to the nucleus that we must turn now to close this chapter.

Like the eukaryotic cell itself, the origin of the nucleus has been the subject of an onslaught of ideas and theories, from simple bubbles of the cell membrane, to the engulfment of whole cells. Most of these ideas fall at the first hurdle. For example, many don't match up to the structure of the nuclear membrane, which is not a continuous sheet, like the outer cell membrane of any cell, but a series of flattened vesicles, riddled with large pores, that are continuous with other internal membranes inside the cell (see Fig. 4.6). Other proposals offer no grounds to explain why any cell would be better off with a nucleus than without it. The standard answer, that the nuclear membrane 'protects' the genes, invites the question 'against what?' Theft? Vandalism? But if there are any universal selective advantages favouring a nucleus, like molecular damage, why did no bacteria ever develop a nucleus? Some of them, as we've seen, have internal membranes that could have served the need.

In the face of little solid evidence, I'd like to raise another gloriously imaginative hypothesis from the ingenious duo we met in Chapter 2, Bill Martin and Eugene Koonin. Their idea has two great merits. It explains why a nucleus should evolve specifically in a chimeric cell, notably one that is half archaea, half bacteria (which, as we've seen, is the most believable origin of the eukaryotic cell itself). And it explains why the nuclei of virtually all eukaryotic cells should be stuffed with DNA coding for nothing, completely unlike bacteria. Even if the idea is wrong, I think it's the *kind* of thing we ought to be looking for, and it still raises a real problem facing the early eukaryotes that has to be solved somehow. This is the sort of idea that adds magic to science, and I hope it is right.

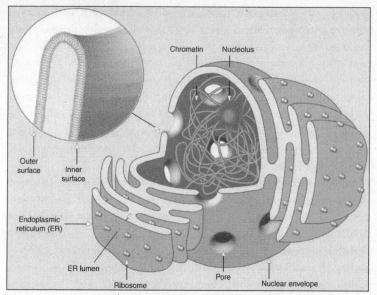

Figure 4.6 Structure of nuclear membrane, which is continuous with other membranes in the cell (specifically the endoplasmic reticulum). The nuclear membrane is formed from these vesicles fusing together. There is no similarity in structure with the external membrane of any cell, implying that the nucleus was not derived from one cell living inside another.

Martin and Koonin considered the curious 'genes in pieces' structure of eukaryotic genes, the discovery of which had come as one of the biggest surprises of twentieth-century biology. Rather than lining up in an orderly fashion like bacterial genes, eukaryotic genes are split up into bits and pieces separated by long non-coding sequences. The non-coding sequences are called *introns* (short for intragenic regions) and their evolutionary history, long perplexing, has but recently emerged blinking into the light.

While there are many differences between introns, we now recognise some shared details that betray their common ancestry as a type of *jumping gene*, able to infect a genome by replicating itself like mad – a selfish gene, out for itself. The trick is simple enough. When a jumping gene is read off into RNA, usually as part of a longer sequence, it spontaneously folds into a shape that amounts to a pair of RNA scissors, and splices itself out of the longer ribbon. It then serves as a template for repeatedly regenerating itself into DNA. The new DNA is incorporated back into the genome, more or less at random, identical copies of the selfish original. There are many types of jumping gene, ingenious variations on a theme. Their astonishing evolutionary success is attested by the human genome project and other big genome sequencing efforts. Almost half the human genome consists of jumping genes or their decayed (mutated) remains. On average, all human genes contain within them three selfish jumping genes, dead or alive.

In a way, a 'dead' jumping gene – one that has decayed to the point that it can no longer jump – is worse than a 'living' one. At least a 'living' jumping gene splices itself out of RNA without doing any real harm; a dead gene just sits there and gets in the way. If it can't splice itself out, the host cell needs to deal with it, otherwise it would get built into a protein and cause mayhem. Eukaryotic cells invented a method of splicing out any unwanted RNA early in their evolution. Interestingly, they simply recruited the RNA scissors from a jumping gene and packed them up with proteins. All living eukaryotes, from plants to fungi to animals, use these ancient scissors to splice out non-coding RNA. So we are faced with a peculiar situation, in which eukaryotic genomes are festooned with introns derived from selfish jumping genes. These are spliced out of the RNA every time a gene is read off using a pair of RNA scissors stolen from the jumping genes themselves. The problem, and the reason why any of this relates to the origin of the nucleus, is that these ancient scissors are a bit slow at cutting.

By and large, prokaryotes don't tolerate either jumping genes or introns. In prokaryotes there is no separation between the genes themselves and the apparatus for building new proteins. In the absence of a nucleus, the protein-building machines, the ribosomes, are mixed up with DNA. Genes are read off into an RNA template that is simultaneously translated into the protein. The problem is that protein building on ribosomes is extremely quick, while the RNA scissors that eliminate introns are slow. By the time the scissors have cut out an intron, the bacterium would have already built several dysfunctional copies of the protein incorporating the intron. Exactly how bacteria rid themselves of jumping genes and introns is not known (purifying selection in large populations could do it), but the fact is that they do. Most bacteria have eliminated nearly all jumping genes and introns although some, including the ancestors of the mitochondria, do have a few. Such bacteria have thirty or so copies per genome, compared with the seething hive of thousands or millions in a eukaryotic genome.

The chimeric ancestor of the eukaryotes apparently succumbed to an invasion of jumping genes from its mitochondria. We know because they look alike, the jumping genes in eukaryotes being similar in structure to the few found in bacteria. What's more, most introns in living eukaryotes are found in exactly the same place within the genes of eukaryotes from amoeba to thistle, from fly to fungus to human. Presumably an early infestation of jumping genes, copying themselves throughout the genome, ultimately 'died' and decayed into fixed introns, in a common ancestor of all the eukaryotes. But why would jumping genes run amok in those earliest eukaryotic cells? One reason is that bacterial jumping genes were hopping around on the chromosome of the host cell, an archaeon, which plainly had no idea how to deal with them. Another reason is that the initial population of chimeric cells must have been small, so the kind of purifying selection that eliminates defects from a large population of bacteria would not have applied.

Whatever the reason, the earliest eukaryotes faced a curious problem. They were infested with introns, many of which would have been incorporated into proteins, because the RNA scissors couldn't splice them out fast enough. While such a situation wouldn't necessarily kill the cells – dysfunctional proteins are broken down and the slow scissors ultimately complete their work to produce functional proteins – it must have been a pretty awful mess. But a solution was staring these muddled cells in the face. According to Martin and Koonin, a simple way to restore order, to produce proper functional proteins all the time, is to make sure that the scissors are given enough time to complete their cutting before the ribosomes get going with the protein-building. In other words, make sure the RNA, with its introns, goes off to the scissors first, and only then is passed on to the ribosomes. Such a separation in time can be achieved simply by a separation in space, by excluding the ribosomes from the vicinity of DNA. With what? With a membrane that has large holes in it! Recruit an existing membrane, hive off your genes within, make sure there are enough pores to export RNA to the ribosomes, and all is well. So the defining feature of all eukaryotes, the nucleus, evolved not to protect the genes at all, according to Martin and Koonin, but to exclude them from the protein-building factories in the cytoplasm.

This solution may sound a little rough and ready (though that, given the way of evolution, is an asset) but it immediately offered some advantages. As soon as jumping genes no longer posed a threat, the introns themselves turned out to be a boon. One reason is that they enabled genes to be cobbled together in different and novel ways, giving a 'mosaic' of potential proteins, a major feature of eukaryotic genes today. If a single gene is composed of five different coding regions, the introns can be spliced out in different ways, giving a range of related proteins from the same gene. While there are only around 25,000 genes in the human genome, these are shuffled about to yield at least 60,000 different proteins, a wealth of variation. If bacteria are the ultimate conservatives, introns turned the eukaryotes into relentless experimentalists.

A second boon is that jumping genes enabled eukaryotes to swell their genomes. Once they had adopted the phagocytic way of life, eukaryotes were no longer bound by the endless drudgery of bacterial life, and specifically the need to streamline themselves for fast replication. Eukaryotes didn't have to compete with bacteria; they could just eat them and digest them within, at their leisure. Freed from the need for speed, those first eukaryotes could accumulate DNA and genes, giving them scope for enormously greater complexity. Jumping genes helped swell eukaryotic genomes up to thousands of times the normal bacterial size. While much of the extra DNA was little more than junk, some was co-opted to form new genes and regulatory sequences. Greater complexity followed almost as a side-effect.

So much for the inevitability of complex life on earth or of human consciousness. The world is split in two. There are the eternal prokaryotes and the kaleidoscopic eukaryotes. The transition from one to the other seems not to have been a gradual evolution, no slow climb to complexity as limitless populations of prokaryotes explored every conceivable variation. Certainly, vast populations of bacteria explored all feasible avenues, but they remained forever bacteria, stymied by their inability to expand in size and energy at the same time. Only a rare and fortuitous event, a collaboration between two prokaryotes, one somehow getting inside the other, broke the deadlock. An accident. The new chimeric cell faced a host of problems, but one great freedom: the liberty to expand in size without incurring a crippling energetic penalty, the freedom to become a phagocyte and break out of the bacterial loop. Faced with an outbreak of selfish genes, a happenstance solution may have given rise not just to the cell nucleus but also to a tendency to collect DNA and to recombine it in the endless constellations of the magical world around us. Another accident. This world of marvels, it seems, springs from two deep accidents. On such tender threads hangs fate. We are lucky to be here at all.

SEX

The Greatest Lottery on Earth

The Irish playwright George Bernard Shaw is a powerful magnet for anecdotes and apocrypha. One such story tells of Shaw being propositioned by a beautiful actress at a party.[1] 'We should have a child together,' the actress declares, 'for it would be blessed with my beauty and your brains.' 'Ah,' replies the cagey Shaw, 'but what if it had my beauty and your brains?'

Shaw had a point: sex is the most peculiar randomiser of successful genes known. Perhaps only the randomising power of sex is able to throw up a Shaw or a beautiful actress in the first place; but no sooner has sex engineered a winning combination of genes than it dissolves them again. An infamous, albeit mostly harmless, organisation, known as the 'Nobel sperm bank', fell into exactly that trap. The biochemist George Wald, on being invited to contribute his prize-winning sperm, declined on the grounds that it was not his sperm they needed but that of people like his father, a poor immigrant tailor, whose loins were unsuspected as the fount of genius. 'What have my sperm given the world?' asked the laureate. 'Two guitarists!' Genius, or intelligence in general, is certainly heritable (which is to say, genes influence rather than determine the outcome), but sex makes it all an unpredictable lottery.

Most of us sense that the magic of sex (as a form of reproduction) lies in exactly this ability to generate variation, to pull a unique being from a hat every time. But when scrutinised with the care of a mathematical geneticist, it is far from obvious that variety for variety's sake is a good thing. Why break up a winning combination; why not just clone it? Cloning a Mozart or a GBS might strike most people as playing God, a dangerous manifestation of humanity's self-inflated conceit, but this is not what the geneticists have in mind. Their point is rather more mundane – the endless variety spun out by sex can lead directly to misery, disease and death, when a plain clone would not. Cloning, by preserving gene combinations fired in the crucible of selection, is often the best bet.

To give a single example, consider sickle-cell anaemia. This is a grave genetic disease, where red blood cells twist into a rigid sickle shape, which can't squeeze through fine capillaries. It is caused by inheriting two 'bad' copies of a gene. Why didn't natural selection eliminate the bad gene, you may ask? Because a single copy of the 'bad' gene is actually beneficial. If we inherit one 'good' and one 'bad' copy from our parents, not only do we *not* suffer sickle-cell anaemia, but we're also less likely to get malaria, another disease that affects the red cells. A single 'bad' copy of the sickle-cell gene alters the membrane of the red cells, blocking the entry of malarial parasites, without turning the cells into a hazardous sickle shape. Only cloning (that is to say, asexual reproduction) can pass on this beneficial 'mixed' genotype every time. Sex shuffles the genes inexorably. Assuming both parents have this mixed genotype, about half of any children do inherit the mixed genotype, but a quarter receive two 'bad' copies of the gene, giving them sickle-cell anaemia, while another quarter end up with two 'good' copies of the gene, putting them at high risk of malaria, at least if they live anywhere in the great swathes of the planet inhabited by the mosquito (which transmits the disease). In other words, greater variety puts no less than half the population at risk of serious disease. Sex can blight lives directly.

And that's far from the only disadvantage to sex. Indeed the list of drawbacks ought to put any sane person off the idea for good. Jared Diamond once wrote a book with the title 'Why is Sex Fun?' but oddly omitted to offer an answer. He must have thought it obvious: if sex were not fun, nobody in their right mind would get up to it. And where would we all be then?

Let's imagine that Shaw threw caution to the wind and gambled his luck on a child with brains and beauty. We'll imagine, too, unfairly but illustratively, that the apocryphal actress lived up to the allegorical reputation of her profession. She probably had venereal disease, let's say syphilis. Their meeting took place before the advent of antibiotics, before syphilis had lost much of its dread among those impoverished soldiers, musicians and artists who frequented equally impoverished ladies of the night. In that age, the dreadful demise into insanity of figures like Nietzsche, Schumann and Schubert made the punishment for sexual misdemeanour all too real. And in those days, the touted cures, like arsenic and mercury, were nearly as bad. A night in the arms of Venus, it was said, led to a lifetime on Mercury.

Syphilis, of course, is just one of many unpleasant or deadly venereal diseases, like AIDS, the incidence of which is now soaring across much of the world. The rise of AIDS in sub-Saharan Africa is shocking and scandalous. As I write, some 24 million Africans are infected with HIV, a prevalence of about 6 per cent among young adults. The worst affected countries have a prevalence of well over 10 per cent, with a related decline in mean lifespan of more than a decade. While the crisis is certainly compounded by inadequate medicine, poverty and co-morbid diseases like tuberculosis, unprotected sex is still the biggest part of the problem.[2] But whatever the cause, the sheer scale of the problem does give a sense of the occupational folly of sex.

But let's go back to Shaw. Unguarded sex with the actress could have produced a child with all the worst traits of its parents, and rendered Shaw himself diseased and insane. But he had some advantages too, unlike all too many of the rest of us. When propositioned by the actress, he was already rich and famous: a magnet not just for anecdotes but, in the modern idiom, for babes. At least by acquiescing to sex he had a chance that some of his genes would flow on down the river of time. He would not have had to endure the wretchedness and torment suffered by so many in the search for the right partner or any mate at all.

I don't want to get into the highly charged politics of sex. It seems obvious that there is a cost to finding a partner, and so a cost to passing on one's genes. I don't mean a financial cost – though that is felt keenly enough by anyone picking up the tab on a first date or staggering away from a divorce settlement – but the cost in unrequited time and emotion, made plain from any lonely-hearts column or the proliferation of internet-dating pages. Yet the real cost, the biological cost, is hard to fathom in human societies, for it is buried under layers of culture and etiquette. If you doubt that there is a serious biological cost, just think of the peacock's tail. Those magnificent plumes, the emblem of male fertility and fitness, are undoubtedly a hazard to survival, as are the colourful courtship displays of plenty of other birds. Perhaps the most striking example of all is the hummingbird. Glorious as they may be, the 3,400 species of hummingbird embody the cost of finding a mate, not for the hummingbird (hard enough, no doubt) but for flowering plants.

Rooted to the spot, plants are the most implausible of sexual organisms, yet the overwhelming majority of them are exactly that; only dandelions, along with a handful of other species, cock a snook at sex. The rest find a way, the most spectacular being the exquisite beauty of flowering plants, which swept through the world some 80 million years ago, turning the dull green forests into the magical painted glades we know today. Although they first evolved in the late Jurassic, perhaps 160 million years ago, their global takeover was long delayed, and ultimately tied to the rise of insect pollinators like bees. Flowers are pure cost to a plant. They must attract pollinators with their flamboyant colours and shapes; produce sweet nectar to make such visits worthwhile (nectar is a quarter sugar by weight); and distribute themselves with finesse – not too close (or inbreeding makes sex pointless) and not too far (or the pollinators will never make it to fertilise a partner). Having settled on a pollinator of choice, the flower and pollinator evolve in tandem, each imposing costs and benefits on the other. And no cost is more extreme than that paid by a tiny hummingbird for the static sex life of plants.

The hummingbird must be tiny, for no larger bird could hover motionless over the deep throat of a flower, its wings humming at 50 beats a second. The combination of tiny size and colossal metabolic rate needed to hover at all means that hummingbirds must refuel almost incessantly. They extract more than half their own weight in nectar every day, visiting hundreds of flowers. If forced to stop feeding for long (more than a couple of hours), they fall unconscious into a coma-like torpor: their heart rate and breathing plunge to a fraction of that in normal sleep, while their core temperature goes into free fall. They have been seduced by the enchanted potions of plants into a life of bondage, moving relentlessly from flower to flower, distributing pollen, or collapsing in a coma and quite possibly dying.

If all that were not bad enough, there is a yet deeper enigma to sex. The cost of finding a partner is as nothing compared to the cost of having a partner at all: the infamous twofold cost of sex. The irate feminist, railing at the very existence of men, has a most reasonable point. On the face of it, men are a heavy cost-indeed, and a woman who solved the problem of virgin birth would be a worthy madonna. While a few men seek to justify their existence by assuming the burden of childcare, or material provisioning, the same is not true of many lower creatures, human or otherwise, where the males quite literally just fuck off. Even so, the impregnated female gives birth to sons and daughters in equal measure. Fifty per cent of her efforts are wasted on bringing ungrateful males into the world, where they simply perpetuate the problem. Any female, in any species without paternal provision, who could do away with males forever, would double her reproductive success. A race of cloning females ought to double in numbers every generation, wiping their sexual relatives from the population in a matter of a few generations. From a purely arithmetical perspective, a single cloning female could swamp a population of a million sexually reproducing individuals in just fifty generations!

Think of this at the level of cells. In clonal reproduction, or virgin birth, one cell divides in two. Sexual reproduction is actually the reverse. One cell (the sperm) fuses with another (the egg) to form a single cell (the fertilised egg). Two cells thus give rise to one: it is replication backwards. The twofold cost of sex manifests itself in gene numbers. Each sex cell, the sperm and the egg, passes on only 50 per cent of the genes of its parent to the next generation. The full quota of genes is re-established when the two sex cells fuse. In this context, an individual that finds a way to pass on 100 per cent of its genes to all its offspring, by cloning, has an inbuilt twofold advantage. Because each clone passes on twice as many genes as a sexual organism, the clone's genes should spread swiftly throughout the population, eventually replacing the genes for sexual reproduction.

It gets worse. Passing on only half your genes to the next generation opens the door to all sorts of other dubious shenanigans with selfish genes.[3] In sex, at least in principle, all genes have a probability of exactly 50 per cent of being passed on to the next generation. In practice, this creates an opportunity for the cheats to do better: to act in their own selfish interests and get passed on to more than 50 per cent of the offspring. This is not just a theoretical possibility that doesn't actually happen. There are many examples of conflict between genes, between parasitic genes that break the law and the law-abiding majority that gang up to stop them. There are parasitic genes that kill the sperm, or even entire offspring that don't inherit them; genes that sterilise males; genes that inactivate their opposite numbers from the other parent; and jumping genes that proliferate throughout the genome. Many genomes, including our own, are stuffed with the relics of jumping genes that once replicated all over the genome, as we saw in Chapter 4. The human genome is a graveyard of dead jumping genes, literally half composed of their decaying corpses. Other genomes are even worse. An unbelievable 98 per cent of the wheat genome is made up of dead jumping genes. In contrast, most organisms that clone themselves have leaner genomes and apparently don't fall prey to parasitic genes in anything like the same way.

All in all, the odds seem massively loaded against sex as a mode of reproduction. An inventive biologist may conceive of peculiar circumstances in which sex could prove beneficial, but most of us, on the face of it, would feel compelled to dismiss sex as an outlandish curiosity. It suffers a notorious twofold cost compared with virgin birth; it propagates selfish genetic parasites that can cripple whole genomes; it places a burden on finding a mate; it transmits the most horrible venereal diseases; and it systematically demolishes all the most successful gene combinations.

And yet despite all that, sex is tantalisingly close to universal among all forms of complex life. Virtually all eukaryotic organisms (those built from cells with a nucleus; see Chapter 4) indulge in sex at some point in their life-cycle, and the large majority of plants and animals are *obligately* sexual, which is to say that we can only reproduce ourselves by sex. This is no quirk. Asexual species, which propagate clonally, are certainly rare, but some, like dandelions, dance under our noses. The surprising fact is that almost all these clones are relatively recent species, typically arising thousands rather than millions of years ago. They are the smallest twigs on the tree of life, and they are doomed. Many species revert to cloning, but they hardly ever reach a mature age in the lifespan of a species: they die out without issue. Only a handful of ancient clones are known, species that evolved tens of millions of years ago and gave rise to large groups of related species. Those that did so, such as the bdelloid rotifers, have become biological celebrities, chaste exceptions in a world obsessed with sex, passing like monks through a red-light district.

If sex is an occupational folly, an existential absurdity, then not having sex is even worse, for it leads in most cases to extinction, a non-existential absurdity. And so there must be big advantages to sex, advantages that overwhelm the foolhardiness of doing so. The advantages are surprisingly hard to gauge and made the evolution of sex the 'queen' of evolutionary problems through much of the twentieth century. It may be that, without sex, large complex forms of life are simply not possible at all: we would all disintegrate in a matter of generations, doomed to decay like the degenerate Y chromosome. Either way, sex makes the difference between a silent and introspective planet, full of dour self-replicating things (I'm reminded of the Ancient Mariner's 'thousand thousand slimy things'), and the explosion of pleasure and glory all around us. A world without sex is a world without the songs of men and women or birds or frogs, without the flamboyant colours of flowers, without gladiatorial contests, poetry, love or rapture. A world without much interest. Sex surely stands proud as one of the greatest inventions of life; but why on earth, how on earth, did it evolve?

✦

Darwin was among the first to ponder the benefits of sex and was pragmatic as always. He saw the principal benefit of sex as hybrid vigour, in which the offspring of two unrelated parents are stronger, healthier and fitter – and less likely to suffer from congenital diseases like haemophilia or Tay–Sachs disease than the children of closely related parents. Examples abounded. One only had to look to the ancient European monarchies like the Hapsburgs, a sickly and insane bunch, to appreciate the ill effects of too much inbreeding. Sex, for Darwin, was all about outbreeding, then, although that didn't stop him marrying his own first cousin, that paragon of virtue Emma Wedgewood, with whom he had ten children.

Darwin's answer had two great merits, but suffers from his total ignorance of genes. The great merits are that hybrid vigour is immediately beneficial and that the benefits are focused on an individual: outbreeding is more likely to produce healthy children, who don't die in childhood, so more of your genes survive to the next generation. This is a nice Darwinian explanation with a wider significance that we'll return to. (Natural selection is operating here on individuals rather than large groups.) The trouble is that this is really only an explanation for outbreeding, not for sex. And so it's not even half the story.

A proper understanding of the mechanics of sex had to wait for decades, until the rediscovery of Austrian monk Gregor Mendel's famous observations on the characters of peas at the beginning of the twentieth century. I must confess that at school I always found Mendel's laws dull to the point of unintelligibility, which I recall with a slight sense of shame. Even so, I do think it's easier to grasp elementary genetics if we skip Mendel's laws altogether, as they were elucidated without any real knowledge of the structure of genes and chromosomes. Let's cut straight to thinking of chromosomes as strings of genes, then, and we can see clearly what is going on in sex, and why Darwin's explanation falls short.

The first step in sex is the fusion of two sex cells – sperm and egg – as we've already seen. Each brings to the union a single set of chromosomes, giving the fertilised egg two complete sets. The two copies are rarely exactly the same, and the 'good' copy can mask the effects of the 'bad' version. This is the basis of hybrid vigour. Inbreeding unmasks hidden diseases because you are more likely to inherit two 'bad' copies of the same gene if your parents are closely related. But this is actually a disadvantage of inbreeding, rather than an advantage of sex. The advantage of hybrid vigour lies in having two slightly different copies of every chromosome that can 'cover' for each other, yet this applies to clones that have two different copies of every chromosome as much as it does to sexual organisms. Hybrid vigour, then, stems from having two different sets of chromosomes, not from sex per se.

It is the second step – the regeneration of sex cells, each with a single copy of every gene – that is the key to sex, and the most difficult to explain. The process is known as *meiosis*, and on the face of it, the division is both elegant and puzzling. Elegant, for the dance of the chromosomes, as they find their partners, clasp them tightly for a time, and then waltz off to opposite poles of the cell, is choreographed with such beauty and precision that the pioneers of microscopy were scarcely able to avert their gaze, concocting dye after dye that captured the chromosomes in the act, like grainy old photographs of an acrobatic dance troupe in their heyday. Puzzling because the steps of the dance are far more elaborate than anyone would have expected from that most utilitarian of choreographers, Mother Nature.

The term *meiosis* comes from the Greek and literally means 'to lessen'. It starts out with a cell that has two copies of each chromosome, and in the end appoints a single copy to each sex cell. This is sensible enough: if sex works by fusing two cells together to forge a new individual with two sets of chromosomes, then it's much easier if the sex cells get one set each. What is startling is that meiosis begins by *duplicating* all the chromosomes, to give four sets per cell. These are then mixed and matched – the technical term is 'recombined' – to generate four entirely new chromosomes, each one taking a bit from here and a bit from there. Recombination is the real heart of sex. What it means is that a gene that once came from your father now finds itself physically sitting on the same chromosome as a gene derived from your mother. The trick might be repeated several times on each chromosome, giving a sequence of genes that runs, for example, *paternal–paternal–maternal– maternal–maternal–paternal–paternal*. The newly formed chromosomes are now unique, different not only from each other but almost certainly from any other chromosome that ever existed (as the crossovers are random, usually in different places). Finally, the cell divides in half, and the daughters divide themselves again, to produce a clutch of four 'granddaughter' cells, each with a single set of unique chromosomes. And that's sex.

It's plain, then, what sex does: it juggles genes into new combinations, combinations that may never have existed before. It does so systematically, across the entire genome. This is equivalent to shuffling a pack of cards, breaking down previous combinations to ensure that all players get a statistically even hand. The question is, why?

The answer that sounds most intuitively reasonable to most biologists, even today, dates back to August Weismann, the ingenious German thinker and heir to Darwin, who argued in 1904 that sex generates greater variation for natural selection to act on. His answer was very different from Darwin's, for it implied the benefit of sex was not to the individual but to the population. Sex, Weismann said, is just as likely to throw up 'good' and 'bad' combinations of genes. While 'good' combinations may be directly beneficial to their owner, 'bad' combinations are just as directly harmful. This means that there is no net advantage or disadvantage to sex for individuals in any generation. Despite this, the population as a whole does improve, Weismann argued, for the bad combinations are weeded out by natural selection, leaving ultimately (after many generations) mostly the good combinations.

Of course, sex in itself doesn't introduce any new variation to a population. Without mutation, sex merely shuffles existing genes around, removing the bad ones as it goes, thereby ultimately restricting variation. Add in a smattering of new mutations to the equation, though, as the great statistical geneticist Ronald Fisher did in 1930, and the advantages of sex get sharper. Because mutations are rare events, Fisher argued, different mutations are most likely to occur in different individuals. In the same way, it's more likely that lightning will strike two different individuals, rather than the same person twice (though both mutations and lightning do sometimes strike the same person twice).

To illustrate Fisher's argument, let's assume that two beneficial mutations arise in a population reproducing clonally. How would they spread? The answer is that they can only spread at the expense of each other, or individuals lacking the mutations (see Fig. 5.1). If both mutations were equally beneficial, the population might end up split fifty-fifty between the two. Crucially, no individual could benefit from both mutations at once unless the second mutation recurred in an individual already benefiting from the first – if the lightning struck twice. Whether that happens frequently, or hardly ever, depends on factors like the mutation rate and the population size; but in general, beneficial mutations will only rarely come together in populations that reproduce in a strictly clonal manner.[4] Sexual reproduction, in contrast, is able to bring together both mutations in a single moment of transcendence. The benefit of sex, then, said Fisher, is that new mutations can be brought together in the

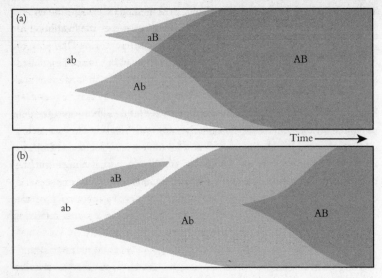

Figure 5.1 The spread of new beneficial mutations in sexual (top) versus asexual (bottom) organisms. In sex, beneficial mutations that convert the gene *a* to *A*, and *b* to *B*, are swiftly recombined to give the best *AB* genotype. Without sex, *A* can only spread at the expense of *B*, or vice versa, so the best genotype *AB* can only be formed when the *B* mutation recurs in an *Ab* population.

same individual almost immediately, giving natural selection the chance to test their combined fitness. If the new mutations do indeed confer greater fitness, sex helps them spread swiftly through the whole population, making organisms better adapted and speeding up evolution (see Fig. 5.1).

The American geneticist Hermann Muller, who received the Nobel Prize for Physiology or Medicine in 1946 for his discovery that X-rays mutate genes, later developed the argument to include detrimental mutations. Having personally generated thousands of mutations in his fruit flies, Muller knew better than anyone that most new mutations are detrimental. For Muller, a deeper philosophical question revolved around how a clonal population could rid itself of such detrimental mutations. Imagine, he said, that almost all the flies had a mutation or two, leaving only a few genetically 'clean' individuals.

What would happen next? In a smallish clonal population, there is no escape from a ratchet-like decline in fitness. The problem is that the likelihood of reproduction depends not only on genetic fitness but also on the play of chance, on being in the right place at the right time. Picture two flies, one of which has two mutations and the other none. The mutant fly happens to find itself in the midst of plentiful food, while the clean fly is starving: despite being less fit, only the mutant survives to pass on its genes. Now imagine that the starved fly was the last of its kind, the only remaining non-mutant: every other fly surviving in the population now has at least one mutation. Unless a mutant fly happens to undergo a back mutation, which is a very rare event, the population as a whole is now one notch less fit than before. The same scenario can be repeated time after time, each time accompanied by a click of the ratchet. Eventually, the entire population will become so degenerate that it falls extinct, a progression now known as Muller's ratchet.

Muller's ratchet depends on chance. If the population is extremely large, then the play of chance is diminished, and the statistical probability is that the fittest individuals will survive; in a large population, the slings and arrows of outrageous fortune cancel out. If the speed of reproduction is faster than the rate at which new mutations accumulate, then the population as a whole is safe from the workings of the ratchet. On the other hand, if the population is small or if the mutation rate is high, then the ratchet clicks into operation. Under such circumstances a clonal population starts to decay, accumulating mutations irrevocably.

Sex saves the day, for sex can recreate unblemished individuals by bringing together unmutated genes in the same individual. In the same way, if two cars are broken – let's say one has a faulty gearbox, and the other a broken engine, using the analogy of John Maynard Smith – then sex is a mechanic who fashions a working car by combining the functional parts. But unlike a sensible mechanic, sex also goes to the trouble of combining the broken parts to produce an unworkable heap of junk. Even-handed as always, the individual benefits of sex are forever cancelled out by individual harm.

There is just one escape clause from this even-handedness of sex, proposed in 1983 by the wily Russian evolutionary geneticist Alexey Kondrashov, now a research professor at the University of Michigan. Kondrashov trained as a zoologist in Moscow, before becoming a theoretician at the Puschino Research Centre, and it was the power of computing that enabled him to arrive at his striking conclusions about sex. His theory makes two bold assumptions, which still provoke rancour among evolutionists. The first is that the mutation rate is rather higher than most people suspected: for Kondrashov's theory to work, there must be one or more deleterious mutations in every individual in every generation. The second assumption is that most organisms are more or less resistant to the effects of single mutations. We only really start to decline in fitness when we inherit a large number of mutations at the same time. This might happen, for example, if the body has some degree of built-in redundancy. Just as we can get by if we lose one kidney, one lung, or even one eye (because the backup organ continues functioning), so at the level of genes there is a degree of overlap in function. More than one gene can do the same thing, buffering the system as a whole against serious damage. If it's really true that genes can 'cover' for each other in this way, then a single mutation wouldn't be too catastrophic, and Kondrashov's theory could work.

How do these two assumptions help? The first assumption – a high mutation rate – means that even infinitely large clonal populations are never safe from Muller's ratchet. They will inevitably decay, ultimately suffering a 'mutational meltdown'. The second assumption is clever. It means that sex can get rid of more than one mutation at once. Mark Ridley, in a delightful analogy, compares cloning and sex to the biblical Old and New Testaments, respectively. Mutations are like sin, says Ridley. If the mutation rate reaches one per generation (everyone is a sinner), then the only way to be rid of sin in a clonal population is to smite the entire population: drown them in a deluge, blast them with fire and brimstone, or infest them with plagues. If, in contrast, sexual organisms can build up a number of mutations without harm (up to a threshold of no return), then sex has the power to amass a large number of mutations in each of two healthy parents and focus them all into a single child. This is the New Testament method. Just as Christ died for collected sins of humanity, so too can sex bring together the accumulated mutations of a population into a single scapegoat, and then crucify it.

The conclusion drawn by Kondrashov is that only sex is able to prevent a mutational meltdown in large complex organisms. The irresistible corollary is that complex life would not be possible without sex. It's an inspiring conclusion, but not one that is generally accepted. Arguments still rage about both of Kondrashov's assumptions, and neither the mutation rate nor the interactions between mutations are easy to measure directly. If there's a consensus at all, it's that the theory might be true in a few instances, but that the assumptions are wrong too often to account for the massive amount of sex going on in the world. Nor does Kondrashov's theory explain the origin of sex among simple unicellular creatures that don't worry about being big and complex, or for that matter about original sin.

Sex, then, benefits populations by bringing together favourable combinations of genes, and by eliminating unfavourable combinations. For the first half of the twentieth century the case was more or less considered closed, although Sir Ronald Fisher did express some reservations about his own theory. In general, Fisher, like Darwin, believed that selection acts on individuals, not for the good of the species as a whole. Yet he felt obliged to make an exception for recombination, which 'could be interpreted as evolved for the specific, rather than the individual, advantage'. While Kondrashov's theory does favour the majority of individuals, only crucifying one every now and then, even in this case the direct benefits of sex can only be felt after many generations. They don't really accrue to individuals, at least not in a conventional sense.

The fuse that Fisher lit was slow-burning, but the time bomb finally exploded in the mid-1960s, as evolutionists began to grapple with ideas of selfish genes and the paradox of altruism. Some of the greatest names in evolutionary theory – George C. Williams, John Maynard Smith, Bill Hamilton, Robert Trivers, Graham Bell, Richard Dawkins – took up the problem. It became clear that little in biology is genuinely altruistic: we are, as Dawkins put it, the blind puppets of selfish genes, which act in their own interests. The question was, from this selfish point of view, why did the cheats not win outright? Why would any individual sacrifice its own best interests now (reproducing clonally) for a benefit (genetic health) that could only accrue to the species at some distant point in the future? Even with all our foresight, humans have a hard time acting in the best interests of our own descendants in the near future – think only of deforestation, global warming and the population explosion. How on earth, then, could blind, selfish evolution place the long-term population benefits of sex over the short-term twofold cost of sex, with all its attendant disadvantages?

One possible answer is that we are stuck with sex because it won't easily 'un-evolve'. If that's the case, then the short-term cost of sex is non-negotiable. As it happens, there is something in this argument. I mentioned earlier that practically all clonal species arose recently, thousands rather than millions of years ago. This is exactly the kind of pattern we would expect if clonal species arise rarely, thrive for a while, and then decay steadily to extinction over thousands of years. Despite the occasional 'flowering' of asexual species, sex is rarely completely displaced because at any one moment there are only a few asexual species around. In fact there are some good 'accidental' reasons why it is difficult for sexual organisms to switch over to cloning. In mammals, for example, a phenomenon known as imprinting (whereby some maternal or paternal genes are switched off) means that any offspring must inherit genes from both parents, or they would be unviable. Presumably, it's mechanistically tricky to reverse such dependence on two sexes; no mammals have given up the sexual habit. Likewise in conifers, two sexes are difficult to expunge because the mitochondria are inherited in the ovule, whereas chloroplasts are inherited in the pollen. To be viable, the offspring must inherit both, requiring two parents. Again, all known conifers are sexual.

But this argument only goes so far. There are several reasons to think that sex does not merely benefit the population, but must have immediate advantages for the individual too. First off, a large number of species – most species, if we consider the vast number of unicellular protists – are *facultatively* sexual, which is to say they only indulge in sex every now and then; even as rarely as every thirty generations or so. In fact some species, like the parasite *Giardia*, have never been caught in flagrante, yet they retain all the genes for meiosis, implying that they might get up to the odd furtive coupling when researchers aren't looking. This logic doesn't only apply to obscure unicellular organisms, but to some large organisms too, like snails, lizards and grasses, which switch from cloning to sex as the circumstances dictate. Obviously, they can revert to cloning whenever they wish, so an 'accidental' block can't be the answer.

A similar argument applies to the origin of sex. When sex was 'invented' by the first eukaryotes (more on this later), there must have been a handful of cells reproducing sexually within a larger population of cells reproducing clonally. To spread throughout the population (as it must have done, because all eukaryotes descend from an ancestor that was already sexual) the act of sex itself must have conferred an advantage on the offspring of sexually reproducing cells. In other words, sex must have spread originally because it benefited individuals within a population, not the population as a whole.

It was this dawning realisation – that sex *must* benefit individuals, despite its twofold cost – that was spelled out by George C. Williams in 1966. The problem had seemed solved, and yet here it was again, now in the most perplexing form. For sex to spread in an asexual population, sexual individuals would have to produce more than twice as many surviving offspring each and every generation. And yet the even-handed mechanics of sex were well understood: for every winner there is a loser, for every good combination of genes there is a bad one. The explanation had to be at once subtle and gigantic, staring us in the face and yet invisible. No wonder it attracted some of the finest minds in biology.

Williams shifted the focus from genes to environment, or more specifically to ecology. Why is it good to be different from your parents, he asked? It could be important if an environment is changing, he answered, or if organisms are invading fresh territory, expanding their niche, dispersing or migrating. Being clonal, Williams concluded, is equivalent to buying a hundred tickets for a lottery, all with the same number. Better to buy fifty tickets, each one with a different number, which is the solution offered by sex.

The idea sounds reasonable, and surely there are instances where it is true; but this was the first of many clever hypotheses to be weighed against the data and found to be wanting. If sex is the answer to a fluctuating environment, then we should find more sex in high latitudes or altitudes, subject to the fickle conditions, or in freshwater streams that flood and dry out in turn. But as a general rule, we don't. There's more sex in stable, highly populated environments, like lakes or the sea, and the tropics, and, in general, if the environment is changing, plants and animals track their preferred conditions, moving north as the climate warms, for example, tailing the receding ice. It's rare, then, for the environment to change so fast that your offspring need to differ *every* generation. Occasional sex really ought to be better. A species that reproduced clonally most of the time, having sex every thirty generations, say, would overcome the twofold cost of sex without losing the benefit of recombination. Yet for the most part that's not what we see, at least among large organisms like plants and animals.

Other ecological ideas, like competition for space, also failed to make headway against the data. But then, from stage left at a run, came the Red Queen. If you've not met her before, the Queen is a surreal character from Lewis Carroll's splendidly absurd work *Through the Looking Glass*. When Alice meets her, she is running furiously, but getting nowhere in particular. 'Now *here*, you see,' the Queen says, 'it takes all the running *you* can do, to keep in the same place.' While biologists use this notion to refer to any arms race between different species locked in perpetual competition – racing against each other but never getting ahead for long – it has most resonance in the evolution of sex.[5]

The Red Queen hypothesis was argued forcefully in the early 1980s by Bill Hamilton, a brilliant mathematical geneticist and naturalist, who many consider to have been the most 'distinguished Darwinian since Darwin'. After making a series of important contributions to Darwinian theory (developing models of kin selection to explain altruistic behaviour, for example), Hamilton now fell under the spell of parasites, tragically falling prey to the malarial parasite at the age of 63, while on a typically fearless expedition to the Congo in 1999 seeking chimpanzees with the AIDS virus. He died in 2000. In a touching obituary for *Nature*, his peer Robert Trivers wrote that Hamilton 'had the most subtle, multilayered mind I have ever encountered. What he said often had double and even triple meanings so that, while the rest of us speak and think in single notes, he thought in chords.'

Before Hamilton's enthusiastic interest, parasites had suffered a bad name. They were dismissed as the contemptible outcome of evolutionary degeneration by the influential Victorian zoologist Ray Lankester (a fate, he believed, that was awaiting Western civilization too); and Lankester's baleful shadow still hung over zoologists a century later. Few researchers outside the discipline of parasitology were willing to countenance the sophisticated adaptations that parasites bring to bear, transfiguring their shape and character from one host to another, homing in on their targets with a miraculous precision, the underpinnings of which eluded parasitologists for decades. Far from being degenerate, parasites are among the most cleverly adapted species known, and more than that, they are fantastically successful – by some estimates outnumbering free-living species by four to one. Hamilton swiftly realised that the unrelenting competition between parasites and their hosts provided exactly the kind of endlessly shifting backdrop needed for sex to offer a big advantage.

Why be different from your parents? Because your parents are probably being lynched from within by parasites, sometimes literally, even as you are being born. Those lucky enough to live in the sterilised conditions of Europe or North America might have forgotten the full horror of parasite infestations, but the rest of the world is not so fortunate. Diseases like malaria, sleeping sickness and river blindness highlight the scale of the misery caused by parasites. Worldwide at least 2 billion people are infected with parasites of one sort or another. We are altogether more likely to succumb to such diseases than to predators, extreme weather conditions or starvation. And more generally, it's not uncommon for tropical animals and plants to host as many as twenty different species of parasite all at once.

Sex helps because parasites evolve rapidly: they have short lifespans and heaving populations. It doesn't take them long to adapt to their host at the most intimate molecular scale – protein to protein, gene to gene. Failure to do so costs them their life; success gives them the freedom to grow and replicate. If the host population is genetically identical, then the successful parasite has the run of the entire population and may well obliterate it. If the hosts vary among themselves, however, there is a chance, indeed a probability, that some individuals will have a rare genotype that happens to resist the parasite. They will thrive until the parasite is obliged to focus its attention on this new genotype or face extinction itself. And so it goes on, generation after generation, cycling genotype after genotype, forever running and getting nowhere, as the Red Queen herself. So sex exists to keep parasites at bay.[6]

Or so the theory says, anyway. It's certainly true that sex is ubiquitous in the highly populated fleshpots where parasites thrive; and it's equally true that sex provides a potentially immediate benefit to individual offspring under these conditions. Even so, there are doubts about whether the threat posed by parasites is really serious enough to explain the evolution and widespread persistence of sex. The kind of relentless genotype cycling predicted by the Red Queen is not easily detected in the wild; and computer models designed to test the conditions fostering sex paint a far more restricted picture than Hamilton's original blazing conception.

In 1994, for example, Curtis Lively, a leading pioneer of the Red Queen hypothesis, admitted that computer simulations showed that 'parasites generated a decisive advantage to sex only when parasite transmission probabilities were high (>70 per cent) and the effects of parasites on host fitness were dire (>80 per cent loss of fitness)'. Although these conditions are undoubtedly true in a few cases, most parasite infections aren't drastic enough to give sex the upper hand. Mutations mean that clones too can become genetically diverse over time, and the computer models show that diverse clones tend to fare better than sexual organisms. Various ingenious refinements do give the Red Queen more power, but smack of special pleading. By the mid-1990s there was no hiding a certain despondency in the field, a sense that no single theory alone could account for the evolution and the persistence of sex.

Of course, there's nothing to say that only one theory has to account for sex. None is mutually exclusive, and although it's a messy solution from a mathematical point of view, nature can be just as messy as she likes. From the mid-1990s, researchers began to combine theories to see if they reinforced each other in some way; and they do. It does matter who the Red Queen shares her bed with, for example, and she is certainly less impotent with some partners. Curtis Lively showed that if the Red Queen and Muller's ratchet are considered together the payback for sex rises, making both ideas more generally applicable. But as researchers went back to the drawing board, looking afresh at different parameters, one leapt out as surely wrong, too mathematical a construct to fit the real world – the assumption of an infinite population size. Not only are most populations far from infinite, but even vast populations are structured geographically, breaking down into assuredly finite, partially isolated units. And that makes a surprising difference.

Perhaps the greatest surprise of all was exactly what it changed. The old ideas about population genetics dating back to Fisher and Muller in the 1930s rose again from the deathly hallows of their resting place in textbooks, to become, I think, the single most promising theory to account for the ubiquity of sex. While a number of researchers from the 1960s onwards have developed Fisher's ideas, especially William Hill, Alan Robertson and Joe Felsenstein, it was the inspiring mathematical treatments of Nick Barton, at the University of Edinburgh, and Sarah Otto, at the University of British Columbia, that really turned the tide. Over the last decade, their modelling has successfully explained sex in terms of the benefit to individuals as well as populations. The new framework also pleasingly incorporates other theories, from Williams's lottery to the Red Queen herself.

The new ideas depend on the interplay of chance and selection in finite populations. In infinite populations, anything that can happen will happen. The ideal combination of genes will emerge inevitably, and it probably won't even take very long. In finite populations, though, the situation is very different. This is because, without recombination, the genes on a chromosome are tied together like beads on a string. The fate of the chromosome depends on the ensemble, on the string as a whole, rather than the quality of individual genes. Most mutations are detrimental, but not so bad that they sink an otherwise good chromosome. This means that they can build up, gradually undermining fitness and contributing ultimately to a poor chromosomal background. Rarely serious enough to maim or kill, the slow drip, drip of mutations saps genetic vigour and imperceptibly lowers the average.

Ironically, when set against this second-rate background, beneficial mutations play havoc. Let's imagine, for the sake of argument, that there are 500 genes on a chromosome. One of two things can happen. Either the spread of the mutation will be retarded by the second-rate company that it keeps on the chromosome, or it won't. In the first case, strong positive selection for one gene is dissipated by weak selection against the other 499. The overall effect is neutral, and there is a good chance that the beneficial mutation will simply be lost again, because the gene is barely visible to natural selection. In other words, interference between genes on the same chromosome, dubbed *selective interference*, obscures the benefits of valuable mutations and obstructs selection.

But the second alternative has sort of a diabolical cunning about it. Imagine there are fifty variants of the same chromosome scattered across a population. A new mutation that's beneficial enough to spread throughout the whole population must, by definition, displace all other variants of the same gene. The trouble is that it doesn't just displace all other variants of the same gene, but all other variants of *all* the genes on the same chromosome in the eclipsed competitors. If the new mutation occurs on one of these fifty chromosomes, the other forty-nine will be lost from the population. In fact, it's even worse than that, for the principle applies not only to genes physically linked on the same chromosome, but to all the genes that share their destiny in a cloned organism, which is to say, *all the genes*. Disastrously, practically all genetic diversity is lost.

Overall then, 'bad' mutations blight 'good' chromosomes, while 'good' mutations get stuck on 'bad' chromosomes, eroding fitness in both cases. In the rare cases where a mutation makes a big difference, strong selection has a disastrous effect on genetic diversity. The outcome can be seen vividly in the degenerate Y chromosome of men, which never recombines.[7] A shadow of the female X chromosome (which does recombine because women have two X chromosomes), the Y chromosome is little more than a stub, with a handful of genes remaining, mixed up with reams of genetic gibberish. If all chromosomes were this degenerate, no forms of complex life would be possible.

And the diabolical cunning doesn't stop there. The stronger the selection, the more likely there will be a selective sweep for one or other gene. Any strong selective force works here, whether parasites or climate, starvation, or dispersal to new habitats, hence the link with the Red Queen and other theories of sex. The outcome in each case is a loss of genetic diversity, which lowers the effective population size. In general, large populations harbour lots of genetic variety, and vice versa. Populations that reproduce clonally will lose genetic variety with every selective purge. From a population-genetics point of view, large populations (in the millions) behave as if they were small (in the thousands) and this opens the door to random chance again. And so heavy selection converts even huge populations to a small 'effective' population size, rendering them vulnerable to degeneration and extinction. A series of studies has shown that exactly this kind of genetic poverty is widespread, not only in clones, but also in species that have sex only sparingly. The great advantage of sex is that it allows good genes to recombine away from the junk residing in their genetic backgrounds, while at once preserving a great deal of the hidden genetic variability in populations.

The mathematical models of Barton and Otto show that 'selective interference' between genes applies to individuals, not just populations. In organisms that can reproduce both sexually and clonally, a single gene can control the amount of sex. The change in prevalence of such a sex gene indicates the success of sex over time. If its prevalence rises, sex wins; if it falls, cloning wins. Crucially, if its prevalence rises from one generation to the next, then sex is benefiting individuals. And rise it does. Of all the ideas that we've discussed in this chapter, selective interference is the most widely applicable. Sex works out better than cloning (despite the twofold cost) under almost any circumstances. The difference is greatest when the population is highly variable, the mutation rate is high, and the selection pressure strong – an unholy trinity that makes the theory conspicuously relevant to the origin of sex itself.

Some of the best minds in biology have wrestled with the problem of sex, but only an incautious minority have been inclined to speculate about its deep origin. There are too many uncertainties about what kind of an entity, or setting, sex evolved in, so any speculation must remain just that. Even so, despite the arguments still raging, there are two statements that I think most in the field could agree upon.

The first is that the common ancestor of all eukaryotes had sex. That's to say, if we reconstruct the shared properties of plants, animals, algae, fungi and protists, one of the chief properties that we all have in common is sex. The fact that sex is so fundamental to eukaryotes is revealing. If we all descend from a sexual ancestor, which in turn descended from asexual bacteria, then there must have been a bottleneck, through which only sexual eukaryotes could squeeze. Presumably the first eukaryotes were asexual like their bacterial forebears (no bacteria have true sex), but all of them fell extinct.

The second point I think everyone could agree about relates to mitochondria, the 'powerhouses' of eukaryotic cells. There's no dispute now that mitochondria were once free-living bacteria, and it seems almost certain that the common ancestor of today's eukaryotes already had mitochondria. There's also no doubt that hundreds if not thousands of genes were transferred from the mitochondria to the host cell and that the 'jumping genes' that festoon the genomes of almost all eukaryotes derive from the mitochondria. None of these observations is especially controversial, but together they paint a striking portrait of the selection pressures that may have given rise to sex in the first place.[8]

Picture it, that first eukaryotic cell, a chimera with tiny bacteria living inside a larger host cell. Every time one of the internalised bacteria dies its genes are set free: they rain on the host cell chromosomes. Fragments of them are incorporated at random into the host's chromosome, by way of the standard bacterial procedures for incorporating genes. Some of these new genes are valuable, others are no use; a few duplicate existing genes. But some are integrated slap in the middle of the host's own genes, chopping them up into bits and pieces. Jumping genes wreak havoc. The host cell has no way of halting their proliferation, so they leap with impunity about the genome, insinuating themselves into genes and slicing the host's circular chromosome into the numerous straight chromosomes that we eukaryotes share today (see Chapter 4).

This is a highly variable population, evolving rapidly. Simple mutations cost the cell its wall. Others are adapting the bacterial cell skeleton into the more dynamic eukaryotic version. The host cell may be forming a nucleus and internal membranes by disorderly transfer of genes for lipid synthesis from its bacterial guests. None of these changes require any hopeful leap into the unknown: all the steps can evolve by simple gene transfers and small mutations. But almost all the alterations are detrimental. For every benefit there are a thousand misguided steps. The only way to forge a chromosome that doesn't kill you, the only way to bring the best innovations and genes together in a single cell, is by sex. Total sex. Not a little half-hearted gene swapping. Only sex can bring together a nuclear membrane from one cell with a dynamic cell skeleton from another, or a protein-targeting mechanism from yet another, and at the same time eliminate all the failures. The randomising power of meiosis might throw up only one winner in a thousand (survivor is a more fitting term) but it's far, far better than cloning. With a variable population, a high mutation rate and a heavy selection pressure (brought about in part by the great barrage of parasitic jumping genes), clones are doomed here. No wonder we all have sex. Without sex, we eukaryotes would never have existed at all.

The question is: if clones were doomed, could sex have evolved fast enough to save the day? The answer, perhaps surprisingly, is 'yes!' Mechanistically speaking, sex could have evolved quite easily. In essence there are three aspects to it: cell fusion, segregation of chromosomes, and recombination. Let's take a quick look at each in turn.

Cell fusion is more or less precluded in bacteria, because the cell wall gets in the way. But lose the wall and the problem may have been the reverse: avoiding fusion. Many simple eukaryotes, such as slime moulds and fungi, fuse together into giant cells with multiple nuclei. Loose networks of cells called syncytia regularly form as part of primitive eukaryotic life cycles. Parasites such as jumping genes, and for that matter mitochondria, profit by fusing cells together, thus gaining access to new hosts. Several have been shown to induce cell fusion. In this context, it might be that the tougher problem was to evolve ways of preventing cell fusion. So the first prerequisite for sex, cell fusion, was almost certainly not a problem.

At first glance, segregation looks more challenging. Recall that meiosis is an intricate dance of chromosomes, which unexpectedly begins by doubling up the number of chromosomes, before segregating a single set into each of four daughter cells. Why so complex? In fact, it's not: it's no more than a modification of the existing method of cell division, mitosis, which also begins by doubling up chromosomes. Mitosis probably evolved via a fairly simple succession of steps from normal bacterial cell division, as articulated by Tom Cavalier-Smith. He goes on to point out that only one key change is necessary to convert mitosis into a primitive form of meiosis – a failure to digest all the 'glue' (technically *cohesin* proteins) holding the chromosomes together. Rather than entering another round of cell replication, duplicating the chromosomes again, the cell pauses for a while, and then continues with the segregation of chromosomes. In effect, the residual glue confuses the cell into thinking that it's primed for the next round of chromosome segregation, before it has completed the first round.

The outcome is a reduction in chromosome numbers, which Cavalier-Smith posits was actually the primary benefit of meiosis in the first place. If the first eukaryotic cells couldn't stop fusing together into networks with multiple chromosomes (as still happens today in slime moulds), then some form of reductive cell division was necessary to regenerate simple cells with a single set of chromosomes. By gumming up the standard method of cell division, meiosis enabled the regeneration of individual cells. It did so with the minimum fuss from an existing mechanism of cell division.

And that brings us to the final facet of sex – recombination. Again, the evolution of recombination doesn't present a problem, as all the machinery needed is present in bacteria, and was simply inherited. Not only the machinery, but the precise method of recombination is exactly the same in bacteria and eukaryotes. Bacteria routinely take up genes from the environment (via a process called lateral gene transfer) and incorporate them into their own chromosome by recombination. In the first eukaryotes, the same machinery must have been responsible for incorporating the bacterial genes that rained down from the mitochondria, giving rise to steadily expanding genome sizes. According to Tibor Vellai, at Eötvös Loránd University, Budapest, the benefit of recombination in the first eukaryotes was probably gene loading, as in bacteria. But to press the recombination machinery into a more general role, in meiosis, was surely a formality.

And so the evolution of sex was probably not difficult. Mechanistically, it was almost bound to happen. The deeper paradox for biologists is that it persisted at all. Natural selection isn't about 'the survival of the fittest', for survival counts for nothing if the fit fail to reproduce. Sex grants a huge head start to cloning, but has prevailed among almost all eukaryotes. The advantage that sex offered at the beginning was probably no different from today

– the ability to bring together the best combinations of genes in the same individual, to purge detrimental mutations, and to incorporate any valuable innovations. In those days, sex might have produced only one winner, or even abject survivor, for every 1,000 failures, but that was still far better than cloning, which must have spelled certain doom. And even today, sex may produce half as many offspring, but ultimately they're more than twice as fit.

Ironically, these ideas date back to the beginning of the twentieth century, falling from favour only to re-emerge in a more sophisticated guise, as more fashionable theories fell by the wayside. They account for sex in terms of benefit to individuals, but also, happily, incorporate older theories in a fashion worthy of sex itself – misguided ideas discarded, fruitful ones accommodated together in a single theory, like genes recombining on a chromosome. For ideas evolve best by sex, too, and we are all the benefactors.

6

MOVEMENT

The Power and the Glory

'Nature, red in tooth and claw' must be one of the most over-quoted references to Darwin in the English language. Even so, the phrase captures the essence, if not of natural selection itself, then at least the popular perception of it. The line is taken from the brooding poem *In Memoriam* by Tennyson, which was finished in 1850, nine years before the publication of Darwin's *Origin of Species*. The poem is a response to the death of Tennyson's friend, the poet Arthur Hallam, and the immediate context of the line is a shockingly bleak contrast between the love of God and the utter indifference of Nature. Not only do individuals perish, Tennyson has Nature say, but so too do species: 'A thousand types are gone: I care for nothing, all shall go!' In our own case 'all' implicitly means all that we hold dear – purpose, love, truth, justice, God. Although never quite losing his faith, Tennyson at times seems to be wracked with doubts.

This stark view of nature, later attributed to the grinding wheels of natural selection, has been attacked from many quarters. Taken literally, the idea at best ignores herbivores, plants, algae, fungi, bacteria, and so on, reducing all life to the vivid struggle between predator and prey. And taken metaphorically, as the more general struggle for existence favoured by Darwin, it tends to play down the importance of cooperation between individuals and species, even of genes within the individual: the importance of symbiosis in nature. I don't want to dwell on cooperation here, but rather to take the line literally and to consider the importance of predation, and specifically the way in which powered movement, or motility, transformed the world in which we live long ago.

'Red in tooth and claw' already implies movement. First catch your prey: not usually a passive quest. But then to clamp your jaws requires opening and closing a mouth with some force: muscles are needed. Claws, too, can hardly tear unless wielded with ferocity, powered by muscles. I suppose if we try to imagine a passive form of predation we might come up with something like a fungus; but even then some form of movement is involved, if only slow strangulation by sucking hyphae. But my real point is that, without motility, predation as a way of life is barely imaginable. Motility, then, is the deeper, the more profound, invention. To capture prey and eat it, you must first learn to move, whether like a tiny amoeba, creeping and engulfing, or with the power, speed and grace of a cheetah.

Motility has indeed transformed life on earth in ways that are not immediately apparent, from the complexity of ecosystems to the pace and direction of evolution among plants. This story is betrayed by the fossil record, which gives an insight, however imperfect, into the webs of interactions between species, and the way in which these change over time. Intriguingly, the fossil record points to a rather abrupt change in complexity following the greatest mass extinction in the history of our planet, that at the end of the Permian period, 250 million years ago, when 95 per cent of all species are thought to have vanished. After this great extinction wiped the slate clean, nothing was ever the same again.

The world was complex enough before the Permian, of course. On land there were giant trees, ferns, scorpions, dragonflies, amphibians, reptiles. The seas were full of trilobites, fish, sharks, ammonites, lampshells, sea lilies (stalked crinoids, almost totally wiped out in the Permian extinction) and corals. A cursory inspection might suggest that some of these 'types' have changed, but that the ecosystems were not so very different; yet a detailed inventory says otherwise.

The complexity of an ecosystem can be estimated by the relative number of species: if a handful of species dominate, and the rest carve out a marginal existence, then the ecosystem is said to be simple. But if large numbers of species coexist together in similar numbers, then the ecosystem is far more complex, with a much wider web of interactions between species. By totting up the number of species living together at any one time in the fossil record, it's possible to come up with an 'index' of complexity, and the results are somewhat surprising. Rather than a gradual accrual of complexity over time, it seems there was a sudden gearshift after the great Permian extinction. Before the extinction, for some 300 million years, marine ecosystems had been split roughly fifty-fifty between the simple and complex; afterwards, complex systems outweighed simple ones by three to one, a stable and persistent change that has lasted another 250 million years to this day. So rather than gradual change there was a sudden switch. Why?

According to palaeontologist Peter Wagner, at the Field Museum of Natural History in Chicago, the answer is the spread of motile organisms. The shift took the oceans from a world that was largely anchored to the spot – lamp-shells, sea lilies, and so on, filtering food for a meagre low-energy living – to a new, more active world, dominated by animals that move around, even if as inchingly as snails, urchins and crabs. Plenty of animals moved around before the extinction, of course, but only afterwards did they become dominant. Why this gearshift took place after the Permian mass extinction is unknown, but might perhaps relate to the greater 'buffering' against the world that comes with a motile lifestyle. If you move around, you often encounter rapidly changing environments, and so you need greater physical resilience. So it could be that the more motile animals had an edge in surviving the drastic environmental changes that accompanied the apocalypse (more on this in Chapter 8). The doomed filter feeders had nothing to cushion them against the blow.

Whatever the reasons, the rise and rise of the motile after the Permian extinction transfigured life. Moving around meant that animals bumped into each other far more often, both literally and figuratively, which in turn enabled a greater web of potential interactions between species: not just more preda-tion, but also more grazing, scavenging and burrowing. There were always good reasons to move, but the new lifestyles that came with motility gave animals a particular reason to be in a particular place at a particular time, and indeed a different place at a different time. That is to say, it gave them purpose – deliberate, goal-directed behaviour.

But the rewards of motility go beyond lifestyle, for motility also dictates the pace of evolution, the rate at which genes, and species, change over evolutionary time. While the fastest evolvers of all are parasites and pathogenic bacteria, which must deal with the endlessly inventive and sadistic persecution of the immune system, animals press them hard. In contrast, filter feeders, and plants in general, fixed as they are to the spot, don't evolve as quickly. The idea of the Red Queen, who must run to stay in the same place, at least in relation to her competitors, is almost alien to a world of fixed filter feeders that remain essentially unchanged for aeons before being wiped out at a stroke. But there is an exception to this rule of thumb, which again emphasises the importance of motility: the flowering plants.

Before the Permian extinction, there weren't any flowering plants to be seen. The plant world was a monotonous green, like a coniferous forest today. The explosion of colour in flowers and fruit was purely a response to the animal world. Flowers, of course, attract pollinators, motile animals, which transport the pollen from one plant to another, spreading wide the benefits of sex for the sessile plants. Fruit, too, calls upon the motility, and guts, of the animal world to disperse seeds. And so the flowering plants have coevolved with the animals, each side in hock to the other – the plants fulfilling the deepest cravings of the pollinators and fruit-eaters, the animals blindly executing the silent stratagems of the plants, at least until we humans started producing seedless fruit. Such interweaving of destinies sped the pace of evolution among the flowering plants to match that of their animal partners.

So motility brings with it a need to deal with rapidly changing environments, more interactions between plants and other animals, new lifestyles like predation, and more complex ecosystems. All these factors encouraged the development of better senses (better ways of 'sampling' the surrounding world) and a faster pace of evolution, simply to keep up, not just among animals, but among many plants too. At the heart of all this innovation is a single invention, which made it all possible: muscle. While not perhaps engendering the same sense of perfection as organs like the eye, when viewed down the microscope muscles are an awesomely purposeful-looking array of fibres acting in concert to exert force. They are machines that convert chemical energy into mechanical force, an invention as fantastical as those of

Leonardo. But how did such a purposeful machine come to be? In this chapter we'll look into the origin and evolution of the molecular machinery that drives the contraction of muscle, which enabled animals to alter the world so utterly.

Few attributes impress quite as much as muscles, and the muscle-bound male has aroused lust or envy, from Achilles to a certain 'Governator' of California. But appearances are not everything. A parallel history has seen some of the greatest thinkers and experimentalists grappling with the question of how muscles actually work. From Aristotle to Descartes, the idea held that muscles didn't so much contract as inflate, like the egos of the muscle-bound. A discharge of invisible and weightless animal spirits surged from the ventricles of the brain through the hollow nerves to the muscles, shortening them as they inflated. Descartes himself, with his mechanistic view of the body, proposed the existence of tiny valves in the muscle, which supposedly prevented the backflow of animal spirits in the same way as valves in blood vessels.

Not long after Descartes, though, in the 1660s, a single finding overturned these long-cherished ideas: the Dutch experimentalist Jan Swammerdam demonstrated that muscle volume does not increase as it contracts, but rather slightly falls. If its volume falls, then the muscle could hardly be swelling with animal spirits like a bladder. Soon afterwards, in the 1670s, another Dutchman, the pioneer of microscopy Antony van Leeuwenhoek, first used his magnificent glass lenses to discern the microscopic structure of flesh. He described slender fibres composed of 'very small conjoyned Globules' strung together in a chain; thousands of chains of these globules made up the structure of flesh. The Englishman William Croone imagined that the globules might act as microscopic bladders, which distended the shape of the muscle without necessarily altering its overall volume.[1] How this actually happened may have been beyond experimental verification, but not the imagination. Several leading scientists proposed that the bladder-filling was literally explosive. John Mayow, for example, suggested that animal spirits were 'nitro-aerial particles'. These, he said, were supplied by nerves, and mixed with sulphurous particles from the blood to produce an explosion analogous to gunpowder.

Such ideas didn't last long. Eight years after his early observations, Leeuwenhoek scrutinised his 'globules' again with a new improved lens and apologised: the flesh fibres were not long lines of tiny bladders at all, but rather they were fibres crossed by regular 'rings and wrinkles', and it was these striations that gave the impression of globules. What's more, by crushing the fibres and viewing their contents under his lens, Leeuwenhoek realised that they, too, were full of yet smaller filaments, a hundred or so in each fibre. Terminology changes, but the segments described by Leeuwenhoek are known today as 'sarcomeres', the filaments within 'myofibrils'. Clearly muscle contraction had nothing to do with inflating bladders, and everything to do with fibres and more fibres.

Even so, despite proposing that motor fibres in muscles could somehow 'glide' over each other, scientists had no real idea of what force compelled them to move. Nearly a century passed before a new power emerged that might conceivably animate all these fibres: electricity.

In the 1780s, Luigi Galvani, professor of anatomy at the University of Bologna, was startled when the leg muscle of a frog contracted violently as he touched it with a scalpel at the same moment that a spark discharged from an electrical machine across the room, even though the frog was dead. Brushing a brass hook with the scalpel during an incision elicited the same response, as did various other circumstances, including an electrical storm. The idea of animation through electricity, soon dubbed galvanism, impressed Mary Shelley, who studied Galvani's reports before writing her gothic novel *Frankenstein* in 1823. In fact Galvani's own nephew, Giovanni Aldini, was something of a prototype. Touring Europe early in the nineteenth century to demonstrate 'galvanic reanimation of the dead', on one famous occasion Aldini electrocuted the severed head of a malefactor before an audience of surgeons, physicians, dukes, and even the Prince of Wales, at the Royal College of Surgeons. When applying his electrical rods to the mouth and ear, Aldini commented, 'The jaw began to quiver, the adjoining muscles were horribly contorted, and the left eye actually opened.'

The physicist Alessandro Volta, at the University of Pavia, was also impressed with Galvani's findings, but disagreed about their cause. Volta insisted that there was nothing electrical about the body itself: galvanism was merely a reaction to external electric charges that were generated by metals. The leg could conduct electricity, he argued, in the same way as brine – it was simply a passive quality. Galvani and Volta embarked on a dispute that was to last a decade, with their passionate supporters splitting into factions in true Italian fashion: animalists versus metallists, physiologists versus physicists, and Bologna versus Pavia.

Galvani was convinced that his 'animal electricity' really did come from within, but struggled to prove it, at least to Volta's satisfaction. The dispute nicely illustrates the power of scepticism to galvanise experimental thinking. In devising experiments to prove his case, Galvani established that muscle is intrinsically *irritable*, as he put it, able to generate a reaction out of all proportion to the stimulus. He even proposed that muscles could generate electricity within themselves by accumulating negative and positive charges across internal surfaces. Current, he said, flowed through pores that opened between the two surfaces.

This was visionary stuff, but sadly Galvani's case also illustrated the power of the victor to write history, even in science.[2] Refusing to swear an oath of allegiance to Napoleon, whose armies were then occupying Italy, Galvani was stripped of his position at the University of Bologna, and died in poverty the following year. His ideas languished for decades, and for a long time he was remembered as little more than the purveyor of occult animalistic forces and as the opponent of Volta. Volta, in contrast, was made a count of Lombardy by Napoleon in 1810, and later had an electrical unit, the *volt*, named after him. Yet, although Volta's name has taken precedence in the history of science for his invention of the first proper battery, the voltaic pile, he was utterly wrong about animal electricity.

Galvani's ideas came to be taken seriously again later in the nineteenth century, most notably by the German school of biophysical research, whose most famous member was the great Hermann von Helmholtz. Not only did this school prove that muscles and nerves really are powered by 'animal' electricity, but Helmholtz even calculated the speed at which an electrical impulse shoots down a nerve, using a method developed by the military for determining the speed of flying canon balls. Nervous transmission turned out to be oddly slow – a few tens of metres per second, rather than the hundreds of kilometres per second of normal electricity – suggesting that there was something different about animal electricity. The difference was soon ascribed to the lumbering movements of charged atoms, or ions, such as potassium, sodium and calcium, rather than the fleeting shifts of will-o'-the-wisp electrons. The passage of ions across a membrane produces a wave of depolarisation, which is to say a transient swing to a more negative charge outside the cell. This affects the nearby membrane, and so surges onwards down the nerve, or within the muscle, as an 'action potential'.

But how exactly did an action potential power muscle contraction? To answer that still required an answer to the bigger underlying question, how do muscles physically contract? Advances in microscopy again pointed to an answer by revealing consistent bands in muscle fibres, thought to correspond to materials of differing densities. From the late 1830s, William Bowman, an English surgeon and anatomist, made a detailed study of the microscopic structure of skeletal muscle of more than 40 animal species, including humans and other mammals, birds, reptiles, amphibians, fish, crustaceans and insects. All were striated into the segments or 'sarcomeres' described 160 years before by Leeuwenhoek. But within each sarcomere, Bowman noticed, were further bands, alternating dark and pale. During contraction the sarcomeres shortened, expunging the lighter bands, giving rise to what Bowman called a 'dark wave of contraction'. He concluded, correctly, that 'contractility resides in the individual segments' (see Fig. 6.1).

Beyond that, though, Bowman backed away from his own findings. He could see that the nerves within muscle didn't interact directly with the sarcomeres at all, so any electrical initiation must be indirect at the very least. And worse, he was worried about smooth muscle, which is found in sphincters and arteries. This lacks the banding of skeletal muscles altogether, yet still contracts perfectly well. In consequence, Bowman felt the bands had little to do with muscle contraction and that the secret of contractility must lie in the invisible structure of molecules, which he thought would remain forever 'far beyond the reach of sense'. He was right about the importance of molecular structure, but wrong about the bands, and indeed the reach of sense. But

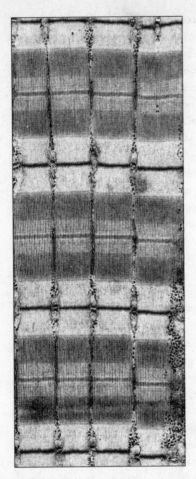

Figure 6.1 The structure of skeletal muscle, showing the characteristic striations and bands. A single sarcomere stretches from one black line (Z disk) to the next. Within a sarcomere, the darkest areas (A bands) contain myosin bound to actin; the light areas (I bands) actin; and the intermediate grey, myosin filaments bound to the M line. When the muscle contracts, the actin in the I band is drawn towards the M line by the myosin cross bridges, shortening the sarcomere and giving rise to a 'dark wave' of contraction (the I band is subsumed by the A band).

Bowman's reservations about muscle structure were shared by almost everyone else.

In a sense, the Victorians knew everything and nothing. They knew that muscles are composed of thousands of fibres, each one divided into segments or sarcomeres, and that these sarcomeres are the basic units of contraction. They knew that the sarcomeres are banded, corresponding to material of differing densities. Some scientists at least suspected that the bands were composed of filaments that slid over each other. They knew too that muscle contractions are electrical, and that electricity is generated as a potential difference across internal surfaces; they even correctly established calcium as the prime suspect. They had isolated the major protein from muscle, and named it *myosin* (derived from the Greek for 'muscle'). But the deeper molecular secrets, which Bowman proclaimed far beyond the reach of sense, were certainly beyond the Victorians. They knew something of the components, but nothing of how they fitted together, nothing of how it all worked. Such insights awaited the virtuosic, reductionist achievements of twentieth-century science. To appreciate the real majesty of muscle, and how the components evolved, we must leave the Victorians far behind and look to the molecules themselves.

Cambridge, 1950: the nascent structural biology unit of the Cavendish Laboratory; a pregnant moment in the history of science. Two physicists and two chemists, grappling with a technology set to transform biology in the second half of the twentieth century: X-ray crystallography. Difficult enough when focused on repetitive geometrical crystals, even today this is a black mathematical art when applied to globular biological molecules.

Max Perutz was head. He and his deputy John Kendrew were the first to determine the structure of large proteins like haemoglobin and myoglobin, from no more than the patterns they produce when a beam of X-rays is scattered by the atoms of their snaking chains.[3] And Francis Crick, soon to be joined by the young American James Watson, famously applied the same technique to the structure of DNA. But in 1950 the fourth man was not Watson but a relatively unknown figure, at least to the outside world, and the only member of that early team who didn't go on to win a Nobel Prize. Yet Hugh Huxley surely should have done, for he more than anyone showed how muscles work at the level of molecular cranks and levers, and his achievements spanned half a century. The Royal Society, at least, honoured him with their highest award, the Copley Medal, in 1997. As I write, he is Professor

Emeritus at Brandeis University, Massachusetts, and still publishing at the age of 83.

Part of Huxley's lesser fame must be put down to unfortunate confusion with his more famous namesake, the Nobel laureate Andrew Huxley, grandson of Darwin's 'bulldog', the ferociously eloquent T. H. Huxley. Andrew Huxley made his name in the post-war years for his studies of nerve conduction, before turning his attention to muscle in the early 1950s; and he, too, was a major figure in muscle research in the following decades. Working independently, the Huxleys – unrelated, to their knowledge – arrived at the same conclusion, and published their results by arrangement back-to-back in *Nature* in 1954. Both proposed what became known as the *sliding filament theory*. Hugh Huxley in particular brought to bear the wonderfully powerful techniques of X-ray crystallography and electron microscopy (only twenty years old at that time). It proved the happiest of combinations, revealing muscle function in finer and finer detail over the following decades.

Hugh Huxley had spent the war working on radar. Returning to finish his degree at Cambridge afterwards, he, like many physicists of his generation, felt compelled by the horrors of the bomb to abandon physics and turn his attention to something less morally and emotionally charged. Physics' loss was biology's gain. Joining Perutz's little group in 1948, he was surprised to find how little biologists knew of muscle structure and function, and embarked on a lifelong quest to put that right. Working on frogs' legs, as had Galvani before him, Huxley's first findings were disappointing. The patterns derived from the muscles of laboratory frogs were faint. But then he found that wild frogs were much better, which signalled a series of long and chilly bike rides over the fens to capture frogs before breakfast. These wild frogs generated X-ray patterns rich in detail but ambiguous in meaning. Ironically, in 1952, Huxley faced Dorothy Hodgkin, one of the pioneers of crystallography, in his PhD examination. On reading his thesis, she leapt to the idea that his data might suggest sliding filaments, and discussed it excitedly with Francis Crick, whom she had bumped into on the stairs. But with the belligerence of youth, Huxley argued, correctly, that she had not read his methods section with proper diligence, and his data did not support her conclusion. Two years later, with the aid of electron microscopic images, Huxley

ultimately came to a similar conclusion himself, but now with proper empiri-
cal support.

But if Huxley had refused to jump to premature conclusions, postponing
his discovery of the sliding filament theory by two years, he was acute in his
early belief that a combination of X-ray crystallography and electron microscopy had the power to unravel the molecular details of muscle contraction.
Both methods were flawed. As Huxley put it: 'Electron microscopy gave one
real tangible images, but with all sorts of artifacts, while X-ray diffraction
gave one true data but in an enigmatic form.' His insight lay in appreciating
that the flaws of one technique could be overcome by the strengths of the
other, and vice versa.

He was lucky, too, for nobody could have foreseen the remarkable advances
over the following half century, most especially in X-ray crystallography.
The difficulty here is with the intensity of the beam. To generate an observ-
able pattern by diffracting (or scattering) X-rays through a structure requires
a large number of rays. This either takes time – hours or even days, back in
the 1950s, when Huxley and others sat overnight cooling weak X-ray sources
– or requires an extremely intense source, capable of producing an intense
beam of X-rays in a fleeting instant. Biologists once again depended on
advances in physics, in particular the development of the synchrotron, those
vast cyclic subatomic particle accelerators, which use synchronised magnetic
and electric fields to accelerate protons or electrons up to astronomical speeds,
before colliding them together. For biologists, the value of the synchrotron
lies in what for physicists is an irritating side-effect. As the particles charge
around in their constrained circles, they release electromagnetic radiation, or
'synchrotron light', much of it in the X-ray range. These fabulously intense
beams can generate diffraction patterns in tiny fractions of a second, patterns
that took hours or days, to generate back in the 1950s. And this was critical,
for the events of muscle contraction take place in hundredths of a second.
Studying the changes in molecular structure taking place during muscle contraction in real time is therefore only feasible using synchrotron light.

When Huxley first put forward the sliding filament theory, it was inevitably
a hypothesis based on rather inadequate data. Since then, though, using
refinements in the same techniques, many of the detailed mechanistic

predictions have been proved, by Huxley and others, measured to atomic resolution over fractions of a second. Where the Victorians could see only rather gross microscopic structures, Huxley could make out the detailed molecular patterns and so postulate mechanism; and today, despite a few remaining uncertainties, we know how muscle contracts almost atom by atom.

Muscle contraction depends on the properties of two molecules, actin and myosin. Both are composed of repeating protein units to form long filaments (polymers). The thick filaments are composed of myosin, already named by the Victorians; the thin filaments, of actin. These two filaments, thick and thin, lie in bundles side-by-side, linked by tiny perpendicular cross-bridges (first visualised by Huxley in the 1950s using electron microscopy). These bridges are not rigid and motionless, but swing, and with each swing they propel the actin filaments along a little, as if the crew of a longboat sweeping their craft through the water. And indeed there is more than an element of the Viking longboat about it, for the oar strokes are unruly, unwilling to conform to a single command. Electron microscopy shows that, of the many thousands of cross-bridges, fewer than half ever pull in unison; the majority are always caught with their oars in disarray. Yet calculations prove that these tiny shifts, even working in disharmony, are together strong enough to account for the overall force of muscle contraction.

All these swinging cross-bridges protrude from the thick filament – they are part of the myosin subunits. On a molecular scale, myosin is huge, eight times larger than an average protein like haemoglobin. In overall shape, a myosin unit is sperm-like; two sperm, in fact, with their heads knocking together and their tails entwined in a frozen embrace. The tails interleave with the tails of adjacent myosin molecules in a staggered array, together making up the heavy filament like the threads of a rope. The heads protrude from this rope in succession, and it is these that form the swinging cross-bridges that interact with the actin filament (see Fig. 6.2).

Here's how the swinging bridges work. The swing-bridge first binds to the actin filament, and once attached, then binds ATP. The ATP provides the

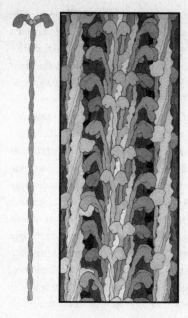

Figure 6.2 Myosin, in the exquisite watercolour of David Goodsell. Left: a single myosin molecule, with its two heads at the top and tails entwined. Right: A thick myosin filament, with heads protruding to interact with actin on either side, and tails entwining like a rope to form the thick filament.

energy needed to power the whole process. As soon as ATP binds, the swing-bridge detaches from the thin filament. The liberated bridge now swings through an angle of about 70 degrees (via a flexible 'neck' region) before binding to the actin filament again. As it does so, the used fragments of ATP are released, and the cross-bridge springs back to its initial conformation, levering the whole thin filament along in its wake. The cycle – release, swing, bind, drag – is equivalent to a rowing stroke, each time moving the thin filament along by a few millionths of a millimetre. ATP is critical. Without it, the head can't release from actin, and can't swing; the result is rigor, as in rigor mortis, when muscles stiffen after death for lack of ATP. (The stiffness fades after a day or so, as the muscle tissue begins to decompose.)

There are many different types of swing-bridge, all broadly similar but differing in their speed. Together, they constitute a 'superfamily' of thousands of members; in humans alone there are about forty distinct types. The speed and force of contraction depends on the speed of the myosin – a fast myosin breaks down ATP rapidly and cranks round the contraction cycle quickly. A number of muscle types are found in every individual and each type has its own myosin, with its own speed of contraction.[4] Similar differences exist between species. The fastest of all myosins are found in the flight muscles of insects such as the fruit fly *Drosophila*, which cycle at a rate of several hundred times a second, nearly an order of magnitude faster than most mammals. As a general rule, small animals have faster myosins; so a mouse's muscles contract at about three times the speed of the equivalent human muscle, a rat's twice as fast. The slowest myosins of all are found in the achingly slow muscles of sloths and tortoises. Here, the speed at which myosin breaks down ATP is around twenty times slower than in a human.

Even though the rate that myosin consumes ATP dictates the speed at which a muscle contracts, running out of ATP never signals the end of muscle contraction. If it did, we would all end gym sessions in a state of rigor, like rigor mortis, and need to be stretchered home each time. Instead, muscles fatigue, which is presumably an adaptation to prevent rigor from setting in. The beginning and end of muscle contractions depends on the level of calcium within the cell, and it is this that connects muscle contraction to Galvani's animal electricity. When an impulse arrives, it spreads swiftly through a network of tubules, which release calcium ions into the cell. Through a number of steps that needn't concern us here, calcium ultimately exposes the sites on the actin filament to which the swing-bridges bind, and this enables the muscle to contract. But no sooner is the muscle cell flooded with calcium than the floodgates are sealed and the pumps spring into action, sucking it all away again in readiness for the next call to action. As the calcium levels fall, the binding sites on the actin filament are covered again, the swing-bridges can no longer bind, and contraction is obliged to cease. The natural elasticity of the sarcomeres ensures that they soon return to their original relaxed state.

This is, of course, a highly simplified account, reducing the working parts to an almost absurd minimum. Consult any textbook and you'll find pages of detail, protein after protein, each with a subtle structural or regulatory role.

Muscle biochemistry is formidably complicated; yet an underlying simplicity shines through. This simplicity is not just an heuristic device: it is central to the evolution of complex systems. In different tissues and species, there are many ways of controlling the binding of myosin to actin. All these various biochemical details are akin to the rococo ornaments of a baroque church; for all that each church might be a masterpiece in its own right, each one is still a baroque church. And likewise, for all the rococo differences in muscle function, myosin always binds to actin, at the same site; and ATP always powers the sliding filaments.

Take smooth muscle, for instance, whose ability to contract sphincters and arteries confounded William Bowman and his Victorian contemporaries. Smooth muscle entirely lacks the striations of skeletal muscle, yet still depends on actin and myosin to contract; the filaments are organised in a far looser manner, shrugging off any sense of microscopic order. The interactions between actin and myosin are also simplified here. A calcium influx activates the myosin heads directly, rather than the roundabout route of skeletal muscle. In other respects, though, the contraction of smooth muscle is similar to skeletal muscles. In both cases, contraction is produced by myosin binding to actin, cranking round the same cycle, powered by the same ATP.

Such relative simplicity might suggest that smooth muscle is a step en route to the evolution of skeletal muscle. Smooth muscle is a contractile tissue that functions well enough, despite lacking a sophisticated microscopic structure. And yet studies of muscle proteins in different species show that muscle evolution wasn't as simple as that. One meticulous study by the geneticists Satoshi OOta and Naruya Saitou, at the National Institute of Genetics, Mishima, Japan, showed that a selection of proteins in the skeletal muscles of mammals are so similar to those in the striated flight muscles of insects that both must have evolved from a common ancestor of vertebrates and invertebrates, living some 600 million years ago. This common ancestor must have had striated muscles, even if it didn't possess a skeleton. Much the same applies to smooth muscle proteins, as they, too, can be traced right back to a similar common ancestor. Smooth muscle was never a step en route to the more complicated striated muscle: it is a separate evolutionary lineage.

That is a remarkable fact. The myosin in our own skeletal muscles is more closely related to the myosin driving the flight muscles of that irritating housefly buzzing around your head than it is to the myosin in the muscles of your own sphincters, tightening in disgust. Astonishingly, the split goes back even further, apparently even predating the origin of bilateral symmetry, shared by both insects and vertebrates. Jellyfish, it seems, also have striated muscles that are minutely comparable with our own. So both smooth muscle and striated muscle contract using a similar system of actin and myosin, but each system apparently evolved independently from a common ancestor that possessed both cell types – a common ancestor numbering among the earliest of animals, from a time when jellyfish were the acme of creation.

Yet despite the unexpectedly long evolutionary separation of striated and smooth muscle, it is plain that all the myriad forms of myosin evolved from a common ancestor. All share the same basic structure, all bind to actin and to ATP at the same sites, all crank through the same motor cycle. If the myosins of both striated and smooth muscle derived from a common ancestor, then that ancestor must have been more primitive than a jellyfish, and probably had neither striated nor smooth muscle, yet still found some use for actin and myosin. What were they doing there? The answer is not new. It dates back to the 1960s, and can be pinpointed to a single unexpected finding. But despite its antiquity, few findings in all of biology have carried such a visual force, while at once opening a dramatic window into the evolution of muscle. Hugh Huxley found that actin can be 'decorated' with myosin heads and viewed by electron microscopy. Let me explain.

All the various filaments can be extracted from muscle and broken down into their component parts. The myosin heads, for example, can be separated from their long tails, and reconstituted with actin in a test tube. The actin swiftly reassembles itself into long filaments; given the right conditions, polymerisation is an inherent property of actin. And then the myosin heads attach themselves to the filaments, just as they do in intact muscle, lining up like little arrowheads all along the filament. All these arrowheads point in the same direction, illustrating the polarity of actin filaments: they always assemble in just one configuration, and myosin always binds in the same direction, hence the generation of force. (In the sarcomeres, this polarity is reversed at the midpoint, drawing both ends towards the middle, thereby contracting

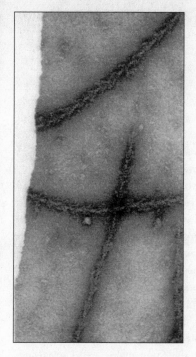

Figure 6.3 Actin filaments derived from the slime mould *Physarum polycephalum*, decorated with myosin 'arrowheads' from rabbit muscle.

each sarcomere as a unit. The contraction of successive sarcomeres shortens the muscle as a whole.)

The little arrowheads bind to actin and nothing else, so adding myosin heads to other cell types works as a test for the presence of actin filaments. Until the 1960s, everyone assumed that actin was a specialised muscle protein, undoubtedly found in the muscles of different species, but not in other types of cell. This conventional wisdom was just being challenged by biochemical data, which suggested that one of the most non-muscular of organisms, brewers' yeast, might contain actin; but the simple expedient of decorating actin with myosin arrowheads opened a Pandora's box of revelations. Huxley opened it first, by adding rabbit myosin to actin filaments extracted from slime mould, a very primitive organism, and finding a perfect match (see Fig. 6.3).

Actin is everywhere. All complex cells contain an internal skeleton, the

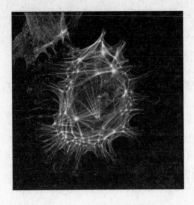

Figure 6.4 Actin cytoskeleton in a cartilage cell from a cow, labelled with the fluorescent dye phalloidin-FITC.

cytoskeleton, fashioned from actin (and other) filaments (see Fig. 6.4). All the cells in our body, as well as all other animals, plants, fungi, algae, protozoa, all have an actin cytoskeleton. And the fact that rabbit myosin binds to slime-mould actin implies that the actin filaments in radically different types of cell are very similar in their detailed structure. This supposition is absolutely correct, startlingly so: we now know, for example, that the gene sequences of yeast and human actin are 95 per cent identical.[5] And from this perspective, the evolution of muscle looks very different. The same filaments that power your muscles power the microscopic world of all complex cells. The only real difference lies in their organisation.

As a musical form, there is something I particularly love about variations. When Beethoven, as a young man, performed for Mozart, it's said that Mozart was not especially impressed with his playing, except his skill at improvisation – his ability to draw out endless rhythmic and melodic variations from a simple theme. Later in life, this skill found its apotheosis in the great Diabelli variations. Like Bach's wonderful Goldberg variations before them, Beethoven's variations are strict in form. The basic harmonic scheme is retained throughout, giving the work as a whole an instantly recognisable unity. After Beethoven, this strictness often lapsed, allowing composers to dwell on moods and impressions, but lacking a mathematical sense of grandeur. Lacking the feeling that every hidden nuance has been teased out, every secret dimension made real, every potential fulfilled.

This ability to take a theme and play out every conceivable variation, while always remaining strictly true to the building blocks of structure, resonates with biology. A set theme, the motor interactions between myosin and actin, for instance, is varied with the endless imagination of natural selection, to arrive at a breathtaking array of form and function. The inner world of any complex cell is testament to this extraordinary facility with strict variation.

The interactions of motor proteins with the filaments of the cytoskeleton are responsible for the whole world of movement in complex cells, both inside and out. Many cells glide along over a solid surface without any apparent effort, no thrashing limbs, no squirming contortions. Others form projections called pseudopodia, which thrust out and drag the cell along, or engulf prey in their protoplasmic folds. Yet others have cilia or flagella, whose sinuous and rhythmic bending drives cells around. Inside cells, the cytoplasm swirls, circulating the contents in an unceasing swell. In this minuscule world of the cell, large objects like mitochondria hasten to and fro, and the chromosomes dance their stately gavotte before withdrawing to their separate corners. And soon afterwards the cell divides in two, constricting itself around the middle with a remorseless corset. All this movement depends upon a molecular toolbox, of which actin and myosin are the exemplary elements. And all of it depends on strict variations of a single theme.

Shrink yourself down to the size of, say, an ATP molecule, and the cell is a vast cityscape of the future. Stretching as far as you can see in all directions is a giddying array of cables, buttressed with yet more cables. Some look flimsy and thin, while others are of great diameter. In this metropolis of the cell, gravity means nothing, viscosity is everything, and the random joggling of atoms is all about. Try to move and you will find yourself stuck as if in treacle, yet at once buffeted and jogged from all sides. And suddenly, through this dizzying city comes a peculiar machine, moving at startling speed, hand-over-mechanical-hand, using one of the cables for a track. Attached to this hurtling machine through an unwieldy coupling device is a truly vast object, towed along at speed. If you were in the way, it would be like being hit by a flying power station. Indeed, that is just what it is – it's a mitochondrion, on its way to power the major works over on the other side of the city. And now that you look, various other objects are all heading in the same direction, some faster, some slower, but all towed along the tracks in the sky by similar machines. And then whoomp! As the mitochondrion passes, you are caught up in the slipstream, whirled along in the vortex. You are part of the constant circulation that stirs the content of all complex cells, the cytoplasmic stream.

This is nanotechnology of a sophistication that we can scarcely begin to imagine; yet for all the strangeness of this futuristic cityscape, there should also be something familiar about it. I could have been describing the inside of one of your own cells, or equally a plant cell, or fungus, or single-celled protozoon swimming around in your local pond. There is a marvellous unity to the world of the cell, which gives a deep sense of connection and fellowship with the world around. From the point of view of the cell, you are just another variation in body plan, just another way of building something wonderful with similar bricks. But what bricks! Each bustling metropolis, common to all eukaryotes (organisms composed of complex cells with a nucleus; see Chapter 4) is very different from the far simpler world inside a bacterium. And much of this difference can be put down to the exuberance of the cytoskeleton and its constant traffic, forever motoring around the contents of the cell. Without this ceaseless traffic flow, the metropolis of the cell would be no more possible than our own great cities without their busy thoroughfares.

All the traffic of the cell is borne by protein motors that work in a broadly similar manner. First is myosin, which cranks up and down the actin filaments, just as it does in muscle. But here the variations begin. In muscle, the myosin heads spend nine tenths of their time detached from the actin filaments; if they didn't, remaining bound instead, they would physically obstruct the other heads from swinging. It would be like being in a boat in which the oarsmen refused to withdraw their oars from the water. In muscle, this arrangement works well because the myosin heads are tethered close to the actin filaments by their long tails, entwined together in the thick filament. But on the high-wire actin filaments criss-crossing the cell, such an arrangement would be more problematic. Once the motors had detached from the filament, they would be buffeted this way and that, and struggle to regain a grip (though in some cases electrical interactions do keep the motor proteins tethered close to their cables).

The best solution here is a 'processive' motor, one that remains attached to the actin wire while somehow moving ('processing') along it. And that is exactly what we get. A few small changes to the structure of myosin turn it into a processive motor, able to move down an actin filament without losing grip. What changes? One is to lengthen the neck. Recall that, in muscle, the two myosin heads knock together, bound tightly by their tails and necks, but otherwise apparently lacking in coordination. Lengthen the necks a little bit, though, and the heads gain a degree of autonomy. One head can remain bound while the other swings, enabling the motor to progress 'hand-over-hand' down the wire.[6] Other variants bring together three or even four heads, rather than just two. And, of course, the tails must go; instead of being wound into a thick filament, the myosin heads are now free to roam. Finally, other objects are attached to the motoring heads via 'coupling' proteins, one for each item of cargo. And there it is: a tribe of processive motors, able to haul cargo in all directions about the cell on the actin tracks.

How did this great parade of motor proteins come to be? There is nothing that compares with it in the world of bacteria. Nor are actin and myosin the only motoring double-act in eukaryotic cells. A second family of motor proteins, called the kinesins, operates in much the same way as the myosins, in a hand-over-hand manner up and down the sky-wires of the cytoskeleton. In the case of the kinesins, though, the sky-wires in question are not the thin actin wires, but higher-bore tubes, known as microtubules, which are assembled from subunits of another protein called tubulin. Among many other tasks, the kinesins are responsible for separating the chromosomes on a spindle of microtubules during cell division. There are other types of motor protein too, but no need for us to get bogged down in detail.

All of these motor proteins, along with their sky-wire tracks, have ancestors in bacteria, though these ancestors are not always obvious, for they usually did quite different jobs.[7] Here again, the technique of X-ray crystallography has revealed kinships that might never have been discerned by gene sequences alone.

At the detailed level of their gene sequences, the two main types of motor protein, the myosins and the kinesins, have virtually nothing in common. Here and there are points of similarity, but for a long time this was taken to be either chance or a case of convergent evolution. Indeed the kinesins and myosins looked to be a classic case of convergent evolution, where two unrelated types of protein became specialised for a similar task, and so developed similarities in structure, just as the wings of bats and birds evolved independently to converge on similar solutions to the common challenge of flight.

But then their three-dimensional structures were solved by crystallography, to an atomic level of resolution. Whereas a gene sequence gives a two-dimensional succession of letters – the libretto without the music – crystallography gives the three-dimensional topography of the protein – the full glorious opera. Wagner once remarked that the music must grow from the words in opera, that the words come first. But nobody remembers Wagner for his heady Teutonic sentiments alone; it is his music that has survived to delight later generations. Likewise the gene sequence is the Word in nature, but the real music of the proteins is hidden in their shapes, and it is the shapes that survive selection. Natural selection cares not a whit about gene sequences; it cares about function. And although genes specify function, it is very often by dictating the shape of the protein, through rules of folding about which we still know little. As a result, various gene sequences can drift so far apart that they no longer bear any resemblance to each other, as in the case of the myosins and kinesins. Yet the deeper music of the protein spheres is still there to be discerned by crystallography.

On the basis of crystallography, then, we know that the myosins and kinesins did indeed share a common ancestor, despite having so little in common in gene sequence. Their three-dimensional shapes show many points of folding and structure in common, right down to critical amino acids being preserved in space with the same orientation. This is an astonishing feat of selection: the same patterns, the same shapes, the same spaces, all are preserved on an atomic level for billions of years, despite the fabric of matter, even the gene sequences, being eroded by the march of time. And these shapes show that both myosins and kinesins are related to a larger family of proteins that plainly have bacterial ancestors.[8] These ancestors performed (and indeed still perform) jobs that involved some form of movement and exertion of force, a switch from one conformation to another, for example, but none of them were capable of true motility. Crystallography, then, shows us the bare bones of protein structure in the same way that X-rays of a bird's wing show us the structure of its skeleton. Just as the ancestry of wings is betrayed by bone structures and joints, which clearly derive from the limbs of flightless reptiles, the structures of motor proteins plainly derive from ancestral proteins capable of conformational change but not true motility.

Crystallography has given tantalising insights into the evolution of the cytoskeleton too, those soaring sky-wires of actin and tubulin. Why, one might ask, would a cell evolve a network of sky-wires, fast tracks for motor proteins, in the absence of those motor proteins? Would that not put the cart before the horses? Not if the cytoskeleton is valuable in its own right. Its value lies in its structural properties. The shape of all eukaryotic cells, from long and spindly neurons to flat endothelial cells, is maintained by the fibres of the cytoskeleton; and it turns out that much the same is true of bacteria. For generations, biologists ascribed many bacterial shapes (rods, spirals, crescents, and so on) to the rigid cell wall bounding the cell, so it came as a surprise in the mid-1990s to discover that bacteria have a cytoskeleton too. This is composed of thin fibres that look a lot like actin and tubulin, fibres that we now know are responsible for maintaining the more elaborate bacterial shapes. (Mutations in the cytoskeleton balloon these complicated bacterial cells back into simple spheres.)

As with motor proteins, there is little genetic resemblance between the bacterial and eukaryotic proteins. Yet the three-dimensional structures, solved by crystallography around the turn of the millennium, are even more striking than the motor proteins. The bacterial and eukaryotic protein structures are virtually superimposable, with the same shapes, the same spaces, and a few of the same critical amino acids in the same places. Plainly the eukaryotic cell skeleton evolved from a similar skeleton in bacteria. With shape, function is preserved. Both serve a broadly structural role; but in both cases, the cell skeleton is capable of more than mere static support. It is not like our solid

bony skeleton, but is dynamic, forever changing, remodelling itself, as incon-
stant and all encompassing as clouds on a stormy day. It can exert force,
moving chromosomes around, dividing cells in half during replication, and
in the case of eukaryotic cells at least, extending cellular projections, without
the help of any motor proteins at all. In short, the cytoskeleton is motile in its
own right. How did such a thing come to be?

Both actin and tubulin filaments are composed of protein subunits that assemble themselves into long chains, or polymers. This ability to polymerise is not
unusual; plastics, after all, are simply polymers composed of basic units
repeated in interminable molecular strings. What is unusual about the
cytoskeleton is that the structure is in dynamic equilibrium: there is an ever-
changing balance between units adding on and others dropping off, polymerisation and depolymerisation. As a result, the cytoskeleton is forever
remodelling itself, building up and breaking down. But here is the magic. The
building blocks can only add on to one end (they fit on to each other like Lego
bricks, or perhaps more exactly, like a pile of shuttlecocks) and dismantle
from the other end of the chain. This gives the cytoskeleton the ability to
generate force. Here's why.

If the rate at which units add on to one end of the chain equals the rate at
which they come off the other, the polymer as a whole maintains a constant
length. In this case, the chain appears to move forwards in the direction that
the subunits are adding on. If an object is caught in the path of the moving
chain, it can be physically pushed along. In fact it's not actually moved along
by the chain itself. What really happens is that the object is buffeted around
by the random joggling of molecular forces; but each time a small gap opens
up between the object and the growing end of the chain, an extra subunit can
wriggle in and bind on. The extension of the chain in this way prevents the
backward movement of the object, and so random joggling tends to push it
forwards.

Perhaps the clearest example of this is seen in some bacterial infections, in
which the bacteria subvert the assembly of the cytoskeleton. *Listeria*, for

example, which can cause meningitis in newborn babies, secretes two or three proteins that together hijack the host cell's cytoskeleton. As a result, the bacteria motor around the inside of the infected cell, pushed by an actin 'comet tail' that associates and dissociates behind them. A similar process is thought to happen in the bacteria themselves, to separate chromosomes and plasmids (small circles of DNA) during cell replication. And again, something similar happens in amoeba (and indeed in some of our own immune cells, like macrophages). The cell's projections, pseudopodia, are pushed out by the dynamic assembly and disassembly of the actin filaments themselves. There's no need for any sophisticated motor proteins at all.

A dynamic cytoskeleton may sound like magic, but according to biochemist Tim Mitchison, at Harvard, that's far from the case. Underlying it all is a spontaneous physical process that takes place without any need for higher evolution. Proteins that have no structural role whatsoever can suddenly polymerise without warning to form a large cell skeleton, capable of generating force, before dismantling equally quickly to return to their former state. Such behaviour might sound alarming, and indeed is usually most unwelcome. In sickle-cell anaemia, for example, a variant form of haemoglobin suddenly polymerises into an internal framework, but only when oxygen levels are low. The change disfigures the red blood cell into the sickle shape that gives the disease its name, in other words, it exerts force and movement. When oxygen levels rise again, this abnormal skeleton dismantles itself equally spontaneously, returning the cell to its normal disc shape. It is a force-generating dynamic cytoskeleton, if not a terribly helpful one.[9]

Something similar must have happened in the case of the cytoskeleton proper, long ago. The units of actin and tubulin fibres are derived from ordinary proteins, with other functions about the cell. A few trifling changes in their structure, as happens with the variant haemoglobin, enabled them to assemble spontaneously into filaments. Unlike sickle-cell anaemia, however, this change must have had an immediate benefit, as it was favoured by natural selection. That immediate benefit might not have been direct or even related to movement. In fact, the sickle-cell haemoglobin is itself selected for in areas where malaria is endemic, as a single copy of the rogue gene protects against malaria. Despite causing prolonged and painful attacks (the sickle cells are inflexible and block capillaries) the spontaneous assembly of an unwanted cell skeleton has been preserved by natural selection because it has an indirect and valuable benefit – keeping out the malarial parasites.

And so the majesty of motility, from its most elementary beginnings, to the many-splendoured power of skeletal muscle, depends on the workings of a handful of proteins, and their endlessly varied forms. The problem that remains today is to tease away all these marvellous variations to expose the original theme, the simple chorale that began it all. This is among the most exciting and disputed fields of research today, for that chorale was sung by the mother of all eukaryotic cells, perhaps as long as 2 billion years ago, and it is hard to reconstruct the echoes of chords so distant in time. How this ancestral eukaryote evolved its motility is not known with certainty. We don't know for sure whether cooperation (symbiosis) between cells played a critical role, as long argued by Lynn Margulis, or whether the cell skeleton evolved from genes already present in the host cell. Some intriguing puzzles, when answered, may shine a brighter light. In bacteria, for example, the chromosomes are drawn apart using actin filaments, whereas the tightening that divides cells during replication is achieved with tubulin microtubules. The reverse is true of eukaryotic cells. Here, the scaffold of the spindle, which separates the chromosomes during cell division, is composed of microtubules, while the contracting corset that divides the cell is made of actin. When we know how and why this role reversal took place, we'll certainly have a better understanding of the detailed history of life on earth.

But these great challenges for researchers are in reality details in an overall scheme that is now broadly clear. We know which proteins the cytoskeleton and motor proteins evolved from, and it matters little in the greater scheme whether they were provided by a symbiotic bacterium or a host cell; each is a plausible source, and when we know the answer, the foundations of modern biology will not collapse. One fact is certain. If there ever were eukaryotes lacking the ability to move around, exerting forces with a dynamic cytoskeleton and motor proteins, then they are no longer to be found: they died out many aeons ago, along with all their progeny. The ancestor of all living eukaryotes was motile. Presumably motility brought with it big advantages. And so it may be that the rise of the motile did more than change the complexity of ecosystems forever. It might have helped change the face of our planet, from a simple world dominated by bacteria to the exuberant world of marvels that we behold today.

7

SIGHT

From the Land of the Blind

Sight is quite a rarity. Eyes are absent, at least in a conventional sense, from the plant kingdom, as well as from the fungi, algae and bacteria. Even in the animal kingdom eyes are not at all common property. There are said to be thirty-eight fundamentally different models of body plan – phyla – in the animal kingdom, yet only six of them ever invented true eyes. The rest have endured for hundreds of millions of years without the benefit of seeing anything at all. Natural selection did not scourge them for lacking sight.

Set against this spartan background, the evolutionary benefits of eyes loom large. All phyla are not equal, and some are far more equal than others. The Chordata, for example, the phylum that includes ourselves and all other vertebrates, comprises more than 40,000 species; the Mollusca, including slugs, snails and octopuses has 100,000; and the Arthropoda, including crustaceans, spiders and insects, numbers more than a million, making up 80 per cent of all described species. In contrast, most of the lesser known phyla, including such oddities as the glass sponges, rotifers, priapulid worms and comb jellies, mostly known only to classically trained zoologists, have relatively few species, tens or hundreds; the Placozoa, just one. If we add them all up, we find that 95 per cent of all animal species have eyes: the handful of phyla that *did* invent eyes utterly dominates animal life today.

Of course, that might be no more than chance. Perhaps there are other subtle advantages to the body plans of these particular phyla that we have missed, quite unrelated to eyes, but that seems unlikely. The evolution of proper eyes, capable of spatial vision rather than simply detecting the presence or absence of light, gives every appearance of having transformed evolution. The first true eyes appeared somewhat abruptly in the fossil record around 540 million years ago, close to the beginning of that 'big bang' of evolution, the so-called Cambrian explosion, when animals burst into the fossil record with breathtaking diversity. In rocks that had been virtually silent for aeons, almost all the modern phyla of animals sprang into existence practically without warning.

The close correspondence in time between the explosion of animal life in the fossil record and the invention of eyes was almost certainly no coincidence, for spatial vision must have placed predators and prey on an entirely different footing; this alone could, and perhaps did, account for the predilection for heavy armour among Cambrian animals, and the much greater likelihood of fossilisation. The biologist Andrew Parker, at the Natural History Museum in London, has made a plausible case that the evolution of eyes drove the Cambrian explosion, in an entertaining, if at times infuriatingly partisan, book. Whether eyes really could have evolved so abruptly (or whether the fossil record is misleading in this regard) is a question we'll consider later. For now let's just note that sight gives far more information about the world than smell, hearing, or touch possibly can, for the earth is drenched in light, and we can hardly avoid being seen. Many of the most marvellous adaptations of life are a response to being seen, whether strutting for sex in the case of a peacock or a flower, parading the great armoured plates of a stegosaurus, or careful concealment in the world of a stick insect. Our own societies are so image-conscious that I scarcely need to labour the point.

Beyond utility, the evolution of sight is culturally iconic, because eyes appear so perfect. From Darwin onwards, eyes have been perceived as an apotheosis, a challenge to the very notion of natural selection. Could something so complex, so perfect, really evolve by unguided means? What possible use, say sceptics, is half an eye? Natural selection calls for a million gradations, each of which must be better than the last, or the half-built structure will be ruthlessly purged from the world. But the eye, say these sceptics, is perfect in the same way as a clock – it is irreducible. Remove a few of the bits and it won't work any more. A clock without hands is worth little, and an eye without a lens or a retina is worthless, or so we're told. And if half an eye is no use then the eye cannot have evolved by natural selection or any other means known to modern biology, and so must be evidence of celestial design instead.

The many vitriolic arguments over perfection in biology rarely do more than entrench already hardened positions. Defenders of Darwin counter that the eye is actually far from perfect, as anyone who wears glasses or contact lenses, or who is losing their sight, knows only too well. This is certainly true, but there is a danger in this kind of theoretical argument, which is to gloss over the many subtleties that undoubtedly exist. Take the human eye, for example. A common argument has it that the design flaws run very deep and are in fact good evidence of the way in which evolution has cobbled together inept unplanned structures, crippled by its own lack of foresight. A human engineer, we're told, would do a much better job; indeed an octopus does. This glib assertion overlooks the mischievous rule known as the second law of Leslie Orgel: Evolution is cleverer than you are.

Let's consider this case briefly. The octopus has an eye much like our own, a 'camera' eye, with a single lens at the front and a light-sensitive sheet, the retina, at the back (equivalent to the film in a camera). Because the last ancestor we shared with the octopus was probably some sort of worm, lacking a proper eye, the octopus eye and our own eye must have evolved independently and converged upon essentially the same solution. This inference is supported by a detailed comparison of the two types of eye. Each develops from different tissues in the embryo and ends up with distinct microscopic organisation. The octopus eye seems to be far more sensibly arranged. The light-sensitive cells of the retina point out towards the light, while the neuronal wires pass back directly to the brain. In comparison, our own retina is often said to be plugged in backwards, an apparently idiotic arrangement. Rather than jutting out, the light-sensitive cells sit at the very back, covered by neuronal wires that pass forwards on a roundabout route to the brain. Light must pass through this forest of wires before it can reach the light-sensitive cells; and worse still, the wires form a bundle that plunges back through the retina as the optic nerve, leaving a blind spot at that point.[1]

But we should not be too quick to dismiss our own arrangement. As so often in biology, the situation is more complex. The wires are colourless, and so don't hinder the passage of light much; and insofar as they do, they may even act as a 'waveguide', directing light vertically on to the light-sensitive cells, making the best use of available photons. And probably more importantly, we have the advantage that our own light-sensitive cells are embedded directly in their support cells (the retinal pigment epithelium) with an excellent blood supply immediately underneath. Such an arrangement supports the continuous turnover of photosensitive pigments. The human retina consumes even more oxygen than the brain, per gram, making it the most energetic organ in the body, so this arrangement is extremely valuable. In all probability the octopus eye could not sustain such a high metabolic rate. But perhaps it doesn't need to. Living underwater, with lower light intensity, the octopus may not need to re-cycle its photopigments so quickly.

My point is that there are advantages and disadvantages to every arrangement in biology, and the outcome is a balance of selective forces that we don't always appreciate. This is the trouble with 'just-so' stories: all too often we see only half the picture. Arguments too conceptual in nature are always vulnerable to counterblasts. Like any scientist, I prefer to follow the train of data. And here the rise of molecular genetics in the last decades furnishes us with a wealth of detail, giving very particular answers to very particular questions. When these answers are all threaded together, a compelling view emerges of how the eye evolved, and from where – a surprisingly remote and green ancestor. In this chapter, we'll follow this thread to see exactly what use is half an eye, how lenses evolved, and where the light-sensitive cells of the retina came from. And in piecing together this story, we'll see that the invention of eyes really did alter the pace and flow of evolution.

It's easy to treat the question 'what use is half an eye?' with derision: which half, the left or the right? I can sympathise with Richard Dawkins's truculent riposte: half an eye is 1 per cent better than 49 per cent of an eye; but for those of us who struggle to conjure up a clear image of half an eye, 49 per cent of an eye is even more stupefying. Actually, though, a literal 'half-an-eye' is a very good way of approaching the problem. The eye does divide neatly into two halves, the front and the back. Anyone who's been to a conference of ophthalmologists will appreciate that they fall into two great tribes: those who work at the front of the eye (cataract and refractive surgeons, dealing with the lens and the cornea), and those who work at the back (the retina), treating such major causes of blindness as macular degeneration. The two tribes interact reluctantly, and at times barely seem to speak the same language. Yet their distinction is a valid one. Stripped of all its optical accoutrements, the eye is reduced to a naked retina: a light-sensitive sheet with nothing on top. And exactly such a naked retina is a fulcrum of evolution.

The idea of a naked retina may sound bizarre, but it fits quite happily into an equally bizarre environment, the deep ocean black-smoker vents that we visited in Chapter 1. Such vents are home to an astonishing array of life, all of which depend, in one way or another, on the bacteria that live directly on the hydrogen sulphide gas emanating from the vents. Perhaps the strangest, and certainly the most celebrated, are the giant tubeworms, which reach eight feet long. Though distantly related to normal earthworms, the tubeworms are literally gutless wonders, possessing neither mouth nor intestine, instead depending for their sustenance on sulphur bacteria nurtured within their own tissues. Other giants found at the vents include huge clams and mussels.

All these giants are found only in the Pacific Ocean, but the Atlantic vents harbour their own wonders, notably the swarming shrimp *Rimicaris exoculata*, which throng in multitudes beneath the smoking chimneys. The name literally means 'eyeless rift shrimp', an unfortunate misnomer that must have returned to haunt its discoverers. Certainly, as might be expected from their name and habitat in the black depths of the ocean, the shrimp don't have conventional eyes. They completely lack the eye-stalks of their surface-dwelling cousins, but they do possess two large flaps on their backs. And although rather nondescript in appearance, these strips reflect light like cats' eyes in the glare of the deep-sea submersibles.

The flaps were originally noticed by Cindy Van Dover, her discovery marking the beginning of one of the more remarkable scientific careers of our times. She is the kind of scientific explorer that Jules Verne used to write about, as endangered a species today as any that she studies. Van Dover now heads the Marine Laboratory at Duke University, and has visited virtually all of the known vents, and more than a few unexplored ones, as the first female pilot of the naval deep-sea submersible *Alvin*. She later discovered that exactly the same giant clams and tubeworms inhabit cold sites on the seafloor, where methane seeps up from the bowels of the earth; clearly the chemical conditions, rather than the heat, are the driving force behind life's exuberance at the bottom of the sea. Back in the late 1980s, though, all this lay ahead, and she must have felt pretty tremulous in sending off samples of the blind shrimp's tissue flaps to a specialist in invertebrate eyes, with the rather lame question, might this be an eye? If you were to mangle a retina, came back the laconic answer, it might look a bit like this. While lacking the usual paraphernalia of eyes – lens, iris, and so on – the blind shrimp possessed what seemed to be naked retinas, running partway down their back, despite living in the black depths of the ocean (see Fig. 7.1).

As more studies were carried out, the findings were better than Van Dover had dared hope. The naked retinas turned out to possess a pigment with properties very similar to that responsible for detecting light in our own retina, called *rhodopsin*. What's more, this pigment was packed into light-sensitive cells characteristic of normal shrimp eyes, even though the overall appearance of the retina was very different. So perhaps the blind shrimp really could see light at the bottom of the ocean. Might the vents themselves produce a faint glow, Van Dover wondered? After all, hot filaments glow and the vents were certainly both hot and full of dissolved metals.

Nobody had ever switched off the lights on *Alvin* before. In pitch blackness, such a manoeuvre was worse than pointless, as there was a good chance that the craft would drift into a vent and fry those aboard, or at least its own instruments. Van Dover had not yet descended to the vents herself, but succeeded in persuading geologist John Delaney, who was about to venture down, to switch off the lights and point a digital camera at a vent. While the blackness was unbroken to the naked eye, Delaney captured on camera a sharply defined halo around the vent, 'hovering in the darkness like the grin of a Cheshire cat'. Even so, these first pictures gave no inkling of what kind of light was emitted – what colour, or how bright. Would it really be possible

Figure 7.1 Eyeless shrimp *Rimicaris exoculata*, showing the two pale naked retinas running down the back.

for the shrimp to 'see' the vents' glow, when we ourselves could see nothing at all?

Like hot filaments, black smokers were predicted to glow red, with wavelengths reaching into the thermal (near infrared) range. In theory, shorter wavelengths, in the yellow, green and blue parts of the spectrum, should not be emitted at all. This prediction was confirmed by some early, albeit crude, measurements, in which coloured filters were placed over the lens. Presumably, if the shrimp could see the vents' glow, then their eyes would need to be 'tuned' to see red or near infrared rays. Yet the first studies of the shrimp's eyes suggested otherwise. On the contrary, the rhodopsin pigment was stimulated most by green light at a wavelength of around 500 nanometres. While this might have been passed off as an aberration, electrical readings of the shrimp retina, though very difficult to carry out, also implied that the shrimp could see only green light. That was odd. If the vents glowed red, and the shrimp could see only green, they were as good as blind. Were these strange naked retinas functionless, then, perhaps degenerate organs like the blind eyes of cave-fish? The fact that they were found on the backs of the shrimp, rather than on their heads, suggested that they were not degenerate but that conjecture hardly amounted to proof.

The proof came with the discovery of larvae. The vent world is not as eternal as it seems, and individual vents often die, choking on their own effluent, in the span of human lives. New vents erupt elsewhere on the ocean floor, often many miles away. For vent species to survive they must cross the void from dying to nascent vents. While the mobility of most adults is hampered by their close adaptations to vent conditions – just think of the giant tube-worms lacking a mouth and gut – their larvae can be disseminated in vast numbers throughout the oceans. Whether the larvae hit on new vents by chance (dispersal by deep ocean currents) or some unknown homing device (following chemical gradients, for example) is a moot point, but the larval forms are not at all adapted to the vent world. For the most part, they are found far closer to the surface, albeit still deep in the sea, at a level where the dying rays of sunlight percolate. In other words, the larvae live in a world where eyes are useful.

Among the first larvae to be identified were those of a crab known as *Bythograea thermydron*. Intriguingly, like the adult vent shrimp, the adult crab flaunts a pair of naked retinas instead of proper eyes; but unlike the shrimp these retinas are found on its head, in the place where one might expect to find eyes. However, the most striking finding was that the larvae of this crab *did* have eyes, perfectly normal eyes, at least for a crab. So when eyes were useful, the crabs had eyes.

There followed a procession of larvae. Several species of vent shrimp live alongside *Rimicaris exoculata*, but are easily overlooked, as they are solitary beasts that don't swarm in such hordes. They too turn out to possess naked retinas, on their heads rather than their backs, and like the crab their larvae have perfectly normal eyes. Indeed, the last larvae to be identified were those of *Rimicaris* itself, partly because the larvae are confoundingly similar to those of other shrimp, and partly because they, too, have quite normal eyes upon their heads.

The discovery of normal eyes on larvae was richly significant. It meant that the naked retina was not merely a degenerate eye – the end-point of generations of loss, any residual function congruent with life in virtual pitch-blackness. The larval forms had perfectly good eyes: if they preferred to lose them during maturation, that had nothing to do with generations of irreversible evolutionary loss: it was something more deliberate, whatever the costs and benefits. By the same token, the naked retina had not evolved 'up' from scratch, attaining a nominal degree of performance that could never rival a true image-forming eye in this benighted environment. Rather, as the larvae mature to the adult form, sinking down to the vents, their eyes degenerate and all but disappear, their fancy optics reabsorbed, step by careful step, leaving but a naked retina. In the case of *R. exoculata* alone, the eyes disappear altogether, and the naked retinas apparently develop from scratch on its back. All in all, a naked retina seems to be more use than a complete eye in a series of different animals: it's not a one-off, not a coincidence. Why?

The value of a naked retina lies in the balance between resolution and sensitivity. Resolution refers to the ability to see (resolve) the details of an image. It improves with a lens, cornea, and so on, as these all help focus light on to the retina, forming an image. Sensitivity is an opposing process, referring to the ability to detect photons. If we have a low sensitivity, we make poor use of available light. In our own case, we can enhance our sensitivity to light by enlarging the aperture (the pupil) and switching to a population of more light-sensitive cells (rod cells). Even so, such measures can only go so far; the mechanical contrivances needed to resolve images at all ultimately restrict our sensitivity. The only way to improve sensitivity any further is to lose the lens and enlarge the aperture indefinitely, increasing the angle through which light can enter the eye. The largest aperture of all is no aperture at all – a naked retina. Taking these factors into consideration, a fairly simple calculation shows that the naked retina of adult vent shrimp is at least 7 million times more sensitive than the fully formed eyes of their own larvae.

So by sacrificing resolution, the shrimp gain the ability to detect extremely low levels of light and, up to a point, where it comes from, at least to the nearest hemisphere: above or below, back or front. Being able to detect light at all could make the difference between life and death in a world poised between temperatures hot enough to cook the shrimp in seconds or too cold and remote to survive. I picture a shrimp drifting off into outer space like an astronaut losing touch with his spaceship. This might explain why *R. exoculata* has eyes on its back, living as it does, in its hordes, on the ledges directly beneath the vents. It is no doubt most comfortable when detecting just the right amount of light filtering down from above on its back, its head buried beneath the thronging multitudes. Its more solitary cousins have apparently forged a slightly different deal, with naked retinas on their heads.

We'll leave the question of why the shrimp see green in a red world for later (they're not colour-blind). For now, the bottom line is that half an eye – a naked retina – is better than a whole eye, at least under certain circumstances. We hardly need to consider how much better half an eye is than no eye at all.

A simple naked retina, a large light-sensitive spot, is in fact the departure point for most discussions of the evolution of the eye. Darwin himself imagined the process beginning with a light-sensitive spot. He's often quoted distressingly out of context on the subject of eyes, not only by those who refuse to accept the reality of natural selection, but even on occasion by scientists eager to 'solve' a problem that supposedly eluded the great man. So he's quoted, correctly, as writing:

> To suppose that the eye, with all its inimitable contrivances for adjusting the focus to different distances, for admitting different amounts of light, and for the correction of spherical or chromatic aberration, could have formed by natural selection, seems, I freely confess, absurd in the highest possible degree.

What is too often omitted is the very next sentence, which makes it plain that Darwin did not consider the eye an obstacle at all:

> Yet reason tells me, that if numerous gradations from a perfect and complex

eye to one very imperfect and simple, each grade being useful to its posses-
sor, can be shown to exist; if, further, the eye does vary ever so slightly, and
the variations be inherited, which is certainly the case; and if any variation
or modification in the organ be ever useful to an animal under changing
conditions of life, then the difficulty of believing that a perfect and complex
eye could be formed by natural selection, though insuperable by our imagination, can hardly be considered real.

In plainer terms, if some eyes are more complex than others, if differences
in eyesight can be inherited, and if poor eyesight is ever a liability, then, says
Darwin, eyes can evolve. All these conditions are fulfilled in plenty. The
world is full of simple and imperfect eyes, from eyespots and pits, lacking a
lens, to rather more sophisticated eyes that parade some or all of Darwin's
'inimitable contrivances'. Certainly eyesight varies, as anyone wearing
glasses, or agonisingly losing their sight, knows all too well. Obviously we're
more likely to be eaten by a tiger or hit by a bus if we don't see it coming. And
of course 'perfection' is relative. An eagle's eyes have four times the resolution of our own, with the ability to detect details a mile away, while we see
some eighty times better than many insects, whose vision is so pixelated that
it could qualify as art.

While I imagine that most people would accept Darwin's conditions
without hesitation, it is still difficult to conceive all the intermediate stages:
imagining the whole continuum, if not actually insuperable, is far from being
superable, to twist P. G. Wodehouse.[2] Unless each step is beneficial in itself, a
complex eye can't evolve, as we've seen. In fact, though, the progression
turns out to be easily superable. It has been modelled as a sequence of simple
steps, shown in Fig. 7.2, by the Swedish scientists Dan-Eric Nilsson and
Susanne Pelger. Each succeeding step is an improvement, beginning with a
naked retina and ending with an eye similar to that of a fish, and not unlike
our own. It could of course (and did) go much further. We could add an iris,
capable of expanding and contracting the pupil to vary the amount of light
entering the eye, from bright sunlight to evening gloom. And we could attach
muscles to the lens to change its shape, pulling and squashing, enabling the
eye to shift its focus from near to distant points (accommodation). But these

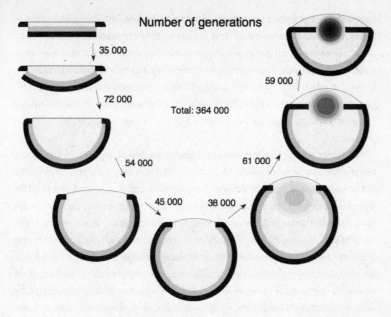

Figure 7.2 The succession of steps needed to evolve an eye, according to Dan-Eric Nilsson and Susanne Pelger, with an approximate number of generations for each change. Assuming each generation is one year, the full progression requires somewhat less than half a million years.

are finessing touches that many eyes lack; and they can only be added to an eye that already exists. So we'll content ourselves with a similar succession in this chapter – to evolve a functional image-forming eye, if still a bit clunky in the optional extras.[3]

The crucial point in this succession is that even the most rudimentary lens is better than no lens at all (anywhere other than a black smoker, of course); a blurred image is better than no image. Again, there is a trade-off between resolution and sensitivity. A perfectly good image can be formed using a pinhole camera, for instance, without a lens at all. Indeed, pinhole eyes exist in some species, notably the nautilus, a living relative of the ammonite.[4] The problem for the nautilus is sensitivity – a sharp image requires a small aperture, so less light can enter the eye. In poor light, the image is so dim as to be practically irresolvable; and that is exactly the problem for nautilus, which lives in deep, darkish waters. Michael Land, one of our foremost authorities on animal eyes at Sussex University, has calculated that adding a lens to an eye of the same size would make it 400 times more sensitive, with 100 times better resolution. So there is a big reward for any steps towards forming a lens of any kind, a reward payable in terms of immediately better survival.

The first 'proper' image-forming lens to evolve probably belonged to a trilobite, whose armour plating is reminiscent of medieval knights, and whose many species held sway in the seas for 300 million years. The oldest trilobite eye peers out from the earliest known trilobite, and dates to 540 million years ago, close to the beginning of the Cambrian 'explosion', as we noted at the start of this chapter. Though modest in comparison with the optical glories attained 30 million years later, the abrupt appearance of trilobite eyes in the fossil record raises the question, could eyes really have evolved so fast? If so, eyesight might well have driven the Cambrian explosion, as argued by Andrew Parker. If not, then eyes must have existed earlier, and for some reason never fossilised; and if that's the case then eyes could hardly have driven the big bang of biology.

Most evidence suggests that the Cambrian explosion happened *when* it did because a change in environmental conditions permitted an escape from the straitjacket of size. The ancestors of the Cambrian animals were almost certainly tiny and lacking in hard parts, explaining the lack of fossils. This would also have prevented the evolution of useful eyes. Spatial vision requires a large lens, an extensive retina, and a brain capable of interpreting the input, and so can only evolve in animals large enough to meet these demands. Much of the groundwork, such as the naked retina and a rudimentary nervous system, was probably in place in small animals living before the Cambrian, but further developments were almost certainly stifled by small size. The immediate impetus for the evolution of large animals was most likely rising levels of oxygen in the air and sea. Large size and predation are only possible in high oxygen levels (nothing else can provide the energy necessary; see Chapter 3) and oxygen rose swiftly to modern levels shortly before the

Cambrian, in the aftermath of a series of global glaciations known as the 'Snowball Earth'. In this electrifying new environment, supercharged by oxygen, large animals living by predation became possible for the first time in the history of the planet.

So far, so good. But if proper eyes didn't exist before the Cambrian, the question resurfaces with even more force, could eyes really evolve so fast by natural selection? There were no eyes at all 544 million years ago and well-developed eyes 4 million years later. On the face of it, the fossils seem to contradict the Darwinian requirement for a million subtle gradations, each one beneficial in its own right. In fact, though, the problem is largely explained by the discrepancy in timescales, between the familiar span of lives and generations, on the one hand, and the numbing passage of geological eras on the other. When measured against the steady rhythm of hundreds of millions of years, any change that happens in a million seems indecently hasty; but it is still an inordinately long time in the lives of organisms. All our modern breeds of dog, for example, evolved from the wolf, admittedly with help from us, in a hundredth of that time.

In geological terms, the Cambrian explosion happened in the blink of an eye – no more than a few million years. In evolutionary terms, though, this is time aplenty: even half a million years ought to have been more than enough time to evolve an eye. In proposing their sequence of steps (see Fig. 7.2), Nilsson and Pelger also computed the time needed. They assumed, conservatively, that each step would be no more than a 1 per cent change to that particular structure, a slightly deeper eyeball, a touch more lens, and so on. When totting up all the steps, they were startled to find that only 400,000 individual changes (not so far from the million I've been glibly throwing around) were needed to progress from a naked retina to a fully formed eye. Then, they assumed that just one change took place per generation (although there could easily have been several at once, making this another conservative estimate). Finally, they assumed that an 'average' marine animal, in which the changes were taking place, would breed once a year. On this basis, they concluded that it would take less than half a million years to evolve a whole eye.[5]

If all these considerations are correct, then the appearance of eyes really could have ignited the Cambrian explosion. And if that's the case, then the evolution of the eye must certainly number among the most dramatic and important events in the whole history of life on earth.

There is one troubling step in Nilsson and Pelger's progression: the first stage in building a lens. Once a primitive lens exists, it's easy to see how natural selection could modify and improve it; but how are the requisite components assembled in the first place? If the various bits and pieces needed to build a lens have no use on their own, should they not be unceremoniously ditched by natural selection before building works can begin? Might this difficulty perhaps explain why the nautilus never developed a lens, even though it would have benefited from doing so?

Actually the question is a non-question, and for now at least the nautilus must remain an oddity for unknown reasons, as most species found a way (including the closest living relatives of the nautilus, octopus and squid); and some found stunningly inventive ways. Although the lens is manifestly a specialised tissue, its construction has been strikingly opportunistic, time and again, its basic building blocks pilfered from any handy nearby source, from minerals and crystals to enzymes, even just bits of the cell.[6]

The trilobite is an excellent example of such opportunism. It really could have transfixed you with a stony eye, for, uniquely, trilobite lenses were made of crystal, the mineral calcite. Calcite is another name for calcium carbonate. Limestone is an impure form; chalk a much purer form. The white cliffs of Dover are almost pure calcite, formed from tiny disorganised crystals that scatter light randomly, giving chalk its white colour. In contrast, if the crystals grow slowly (often in mineral veins) calcite can form into fine, clear structures with a slightly wonky cube shape, known as rhombs. Rhombs have a curious optical property, which arises naturally from the geometry of the constituent atoms: they deflect light from all angles, except one privileged axis straight across the middle. If light enters a rhomb along this axis, dubbed the c-axis, it passes straight through, unhindered, as if ushered along a red carpet. This curious property the trilobites turned to their advantage. Each one of the many eye facets wields its own tiny calcite lens along this privileged c-axis (see Fig. 7.3). Light can pass through each lens from only one direction to the underlying retina.

Quite how the trilobites grew crystal lenses, aligning whole arrays with the correct orientation, is unknown and is likely to remain so, as the last one expired in the Permian extinction, 250 million years ago. But the fact that trilobites are silenced by this vast tract of time doesn't mean that there is no way of knowing how it might have come about. One good clue cropped up from an unexpected source in 2001. It seems that the trilobite lens is not as unique as once thought: one living animal, a brittlestar, also makes use of calcite lenses to see.

There are around 2,000 species of brittlestar, each sprouting five arms like their starfish cousins. Unlike starfish, the brittlestars have long, slender arms hanging down, which snap off if pulled upwards (hence their name). All brittlestars have skeletons made from interlocking calcite plates, which also form spikes on their arms used for grasping prey. Most of them are insensitive to light, but one species, *Ophiocoma wendtii*, confounds observers by scuttling for dark crevices a metre away at the approach of a predator. The trouble is it doesn't have eyes or, at least, so everyone thought until a research team from Bell Labs noticed arrays of calcite knobs on its arms, resembling the lenses of trilobites (see Fig. 7.4). They went on to show that these knobs do indeed function as lenses, focusing light on to photosensitive cells below the lenses.[7] Though lacking anything we'd recognise as a brain, the brittlestar has functional eyes. As *National Geographic* had it at the time: 'In a twist of nature, the sea has eyes in its stars.'

How do the brittlestar lenses form? Although many details are still missing, in broad terms they form in the same way as other mineralised biological structures, like the spines of sea urchins (also made of calcite). The process begins within cells, where high concentrations of calcium ions interact with proteins, which bind them into fixed positions, 'seeding' crystal formation in the same way that an optimist, standing outside an empty grocery store, used to seed a queue in the Soviet Union. One person, or one atom, is immobilised, and the rest stick to that.

In a celebration of reductionism, if the proteins responsible for seeding the calcite crystals are purified and smeared on a sheet of paper, then placed

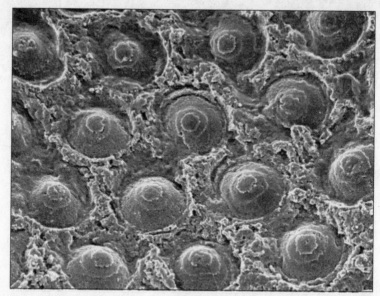

Figure 7.3 Trilobite crystal lenses from *Dalmanitina socialis*, found in Ordovician rocks in Bohemia, Czech Republic; showing details of the inner surface of the lenses; approximately half a millimetre across.

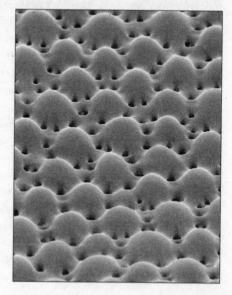

Figure 7.4 The crystal lenses of the brittlestar *Ophiocoma wendtii*, which are found on the skeletal plates at the top of each arm, protecting the joints.

Figure 7.5 Rhombic crystals of calcite, growing on paper smeared with acidic proteins from mollusc shells, placed in a strong calcium carbonate solution. The optical *c*-axis, the only direction in which light passes through the crystal without scattering, is pointing straight up.

in a strong solution of calcium carbonate, perfect crystals grow right there on the paper, forming into rhombs with their optical *c*-axis pointing straight up, just as in a trilobite lens (see Fig. 7.5). There's even a hint of how it came about in the first place: the exact choice of protein doesn't matter much. It matters only that the protein should bristle with acid side-chains. Back in 1992, a decade before anyone knew about the brittlestar lenses, the biomineralologists Lia Addadi and Stephen Weiner grew lovely calcite lenses on a sheet of paper, using acid proteins isolated from mollusc shells, which certainly can't see. In other words, marvellous as it all is, the entire process takes place spontaneously when common proteins mix with common minerals. It is marvellous, certainly, but no more miraculous than the fantastic bristling arrays of crystals found in natural caverns like the Cave of the Swords in Mexico.

Yet for all their sharp sight, crystal eyes were a blind alley. The real significance of trilobite eyes lies in their historical importance, as the first true eyes, rather than in their being a lasting monument of evolution. Other types of natural crystal have been pressed into service by other creatures, notably guanine (one of the building blocks of DNA), which crystallises into sheets that can focus light. Guanine crystals give fish scales their silvery iridescence, and have been added to many cosmetics for the same reason; they're also found in (and take their name from) guano, the dried excrement of birds and bats. Similar organic crystals function as biological mirrors; they're familiar to most people as the 'reflectors' in cats' eyes. They improve night vision by bouncing light back into the retina, giving the receptors a second chance to capture the few scarce photons. Other mirrors focus light on to the retina to form an image. These include the beautiful and numerous eyes of scallops, which peep out from between the tentacles at the rim of the shell; they use a concave mirror beneath the retina to focus light. The compound eyes of many crustaceans, including prawns, shrimps and lobsters, also rely on mirrors to focus light, again using natural crystals along the lines of guanine.

In general, though, the central thrust, and the greatest glories, of evolution have been lenses composed of specialised proteins, like our own. Are these, too, opportunistic constructions, cobbled together from existing components that have other uses about the body? Although it's sometimes claimed that evolution is a historical science, and so cannot be proved one way or another, it does in fact make very specific predictions that can be tested. In this case, the theory predicts that the lens proteins should be recruited from among existing proteins, with other uses about the body, on the grounds that a specialised lens protein could not possibly evolve before the existence of the lens itself.

The human lens is obviously an extremely specialised tissue: it is transparent; blood vessels are excluded; and the cells have lost almost all their normal features, instead concentrating proteins into a liquid crystal array, capable of bending light to form a clear image on the retina. And, of course, the lens is capable of changing shape to alter the depth of field. What's more, the extent to which light is bent varies across the lens, evading faults like spherical aberration (where light passing through the centre or the edge of the lens is focused at different points). Given all this we might conjecture that the proteins needed to fabricate such a refined array would be unique, with optical properties simply not found in mundane everyday proteins. But if we did, we'd be totally wrong.

The proteins found in the human lens are called crystallins, named in the expectation that they would indeed possess unique properties. They make up some 90 per cent of all proteins in the lens. Because lenses are so similar in different species, in both their appearance and function, it seemed reasonable to assume that all were composed of a similar protein. Yet when techniques for comparing the sequence of building blocks in different proteins became widely available, from the early 1980s, the reality came as a surprise. The crystallins are *not* structural proteins and most of them are not even specific to the lens; all do other jobs elsewhere about the body. Even more unexpectedly, many crystallins turned out to be enzymes (biological catalysts) with normal 'housekeeping' functions somewhere else in the body. The most abundant crystallin in the human eye is called α-crystallin, for instance; it is related to a stress protein first found in the fruit fly *Drosophila*, and now known to be widespread in animals. In humans, it functions as a 'chaperone', which is to say, it shields other proteins from damage. As such, it's not only found in the eye, but also in the brain, liver, lung, spleen, skin and small intestine.

To date, eleven types of crystallin have been catalogued, only three of which are found in the eyes of all vertebrates; the rest vary from one group to another, implying they have been 'recruited' to work in the lens quite independently, again exactly as predicted for the ad hoc approach of natural selection. We won't dwell on the names or functions of these enzymes, but it's shocking that this group of metabolic proteins, each with its own task about the cell, has been plucked out and pressed into an utterly different service. It's as if the army conscripted only tradesmen, and members of guilds at that, to form the standing army. But whatever the reasons, there is nothing about this eccentric policy to suggest that it's particularly hard to recruit lens proteins.

All in all, there is nothing special about lens proteins: they are plucked from elsewhere in the body and pressed into service. Virtually all proteins are transparent, so colour is not an issue (only proteins coupled to pigments, like haemoglobin, have much colour of their own). Changes in optical properties like the degree of light-bending (refraction) across the lens are achieved simply by varying protein concentration, which certainly requires finesse, yet is hardly a big conceptual stumbling block. Why so many lens proteins are enzymes, if there is any reason at all, is unknown; but whatever the reason, perfectly formed lens proteins plainly did not leap from the head of Zeus.

A window into how all this comes about is the lowly invertebrate known by the undignified name *sea squirt* (specifically *Ciona intestinalis*, or literally 'pillar of intestines'; Linnaeus was hardly any more kindly). The adult form gives away little of its heritage, being essentially a translucent bag attached to a rock, bearing two yellowish swaying siphons, through which water enters and exits. They are so common in coastal waters around the UK that they're considered a pest. But the larvae divulge the squirt's deep secret, and show them to be much more than just a pest. The larvae look a little like tadpoles and can swim around, making use of a rudimentary nervous system and a pair of primitive eyes, which lack lenses. Once the little squirt has found a suitable home, it attaches itself soundly to the spot and then, needing it no longer, reabsorbs its own brain (a feat that arouses much admiration among university professors, Steve Jones quips).

Though the adult sea squirt is unrecognisable as anything related to ourselves, the tadpole larvae give the game away: the sea squirt is a primitive chordate, which is to say it has a notochord, a forerunner of the spinal chord. This places it among the earliest branches of the chordates, and so all vertebrates. In fact it split from the vertebrates before the evolution of the lens. And this means that the sea squirt, with its simple eye, might give an insight into how the vertebrate lens formed in the first place.

And so it does. In 2005, Sebastian Shimeld and his colleagues in Oxford found that, despite lacking a lens, *C. intestinalis* does have a perfectly good crystallin protein, not in its eye, but tucked away in its brain. Who knows what it's doing there, but that's not relevant to us here. What is relevant is that the same genes that direct lens formation in vertebrates also control the activity of this protein; in the squirt they function in the brain as well as in the eye. So the entire apparatus for building a lens was present in the common ancestor of the vertebrates and sea squirts, before each went its own way. A small switch in regulation in vertebrates transferred the protein from brain to eye. Presumably, similar poaching forays account for the repeated conscription of other crystallins from elsewhere in the body, some in the common ancestor of vertebrates, and others more recently in specific groups. Why the sea squirt line failed to make a simple switch in resource use is a mystery; perhaps a rock is not a hard place to find, even without a lens. Even so, the sea squirt is the odd one out. Most vertebrates did succeed – it happened at least eleven times. And so there are no particularly difficult steps in the sequence to make an eye.

From this riot of appropriated proteins, crystals and minerals that make up the lens of diverse species, the proteins of the retina stand in stark contrast. One protein in particular stands out: the molecule responsible for sensing light, rhodopsin. Recall the vent shrimp *Rimicaris exoculata*, with its naked retinas. Despite the utter peculiarity of the deep ocean vent world, despite the strangeness of the naked retinas down its back, despite the shrimp's ability to detect a faint glow where we cannot, despite living on sulphur bacteria, having blue blood and lacking a backbone, despite last sharing a common ancestor with us around 600 million years ago, long before the Cambrian explosion, despite all this, the vent shrimp use the same protein that we do to see. Is this deep link across time and space no more than an uncanny coincidence, or is it something more significant?

The shrimp's protein and ours are not exactly the same, but they're so similar that if you turned up in court and tried to convince a judge that your version was not a badly concealed plagiarism, you'd be very unlikely to win. In fact, you'd be a laughing stock, for rhodopsin is not restricted to vent shrimp and humans but is omnipresent throughout the animal kingdom. We know little about the inner workings of the trilobite eye, for example, which were not preserved along with its crystal lenses; but we know enough about its relatives to say with some certainty that its eyes would have contained rhodopsin. Every animal, with remarkably few exceptions, relies on exactly the same protein. Trying to persuade a judge that your rhodopsin is not plagiarised would be like trying to claim that your television set is fundamentally different from everyone else's, just because it's bigger or has a flat screen.

This remarkable conformity could conceivably have come about in several ways. It could mean that everyone inherited the same protein from a common ancestor. There's been a lot of little changes in the last 600 million years, of course, but it's still obviously the same protein. Or it could mean that there are such serious design constraints on molecules able to detect light at all that everyone has been forced to come up with basically the same thing. That would be like watching television on a computer screen, a case of different technologies converging on a similar solution. Or, finally, it could mean that the molecule has been passed around freely from one species to another, in a case of rampant theft rather than inheritance.

It's easy enough to discard the third option. Gene theft does exist between species (genes are moved around by viral infections, for example) but it's not common outside bacteria; and when it happens it sticks out like a sore thumb. The catalogue of trifling differences between proteins across species can be superimposed over the known relationships between the species. If the human protein happened to be stolen and inserted into a vent shrimp, it would stare back at us like an illegal alien, clearly related to humans rather than shrimp. On the other hand, if the differences had gradually accumulated over time in the ancestors of the shrimp, then the shrimp protein would be most similar to its close relatives, the prawns and lobsters, and would be most different from its most distant relatives, like us; and this is indeed the case.

If it wasn't stolen, was rhodopsin reinvented from engineering necessity? This is harder to say for sure as there is indeed a sense of reinvention, if only once. The vent shrimp's rhodopsin is about as distant from our own as it's possible to get, for two very similar molecules. In between the two is a spectrum of intermediates, but this spectrum is not quite continuous. Instead, it falls into two groups, roughly corresponding to the vertebrates on the one hand and the invertebrates on the other (including the shrimp). This difference is magnified by a whole context of opposites. In both cases, the light-sensitive cells are modified nerve cells, but there the resemblance ends. In shrimp and other invertebrates, the rhodopsin is plugged into membranes that sprout from the top of the cell like spiky hair (microvilli); in vertebrates, a single projection (a cilium) protrudes from the top of the cell like a radio mast. This mast is convoluted into a succession of deep horizontal folds, making it look more like a stack of discs sitting on top of the cell.

Inside the photoreceptor cells, these differences have their counterparts in biochemistry. In the vertebrates, when light is absorbed, a cascade of signals strengthens the electric charge across the membrane of the cell. The invertebrates do exactly the opposite: when light is absorbed, a completely different cascade causes the membrane to lose its electric charge altogether; and it is this that triggers the nerve to fire off its message *light!* to the brain. All in all, two rather similar rhodopsins are found in utterly contrasting cell types. Does all this mean that the photoreceptor cells evolved twice, once in the invertebrates, and again in vertebrates?

That certainly sounds like a plausible answer and was exactly what most of the field believed until the mid-1990s, when suddenly everything changed. None of the facts is wrong; it's just that they turned out to be only half the story. Now it looks as if everyone uses rhodopsin because everyone inherited it from a common ancestor. It looks as if the earliest precursor of the eye only evolved once.

The iconoclastic Swiss developmental biologist Walter Gehring, at the University of Basel, has promulgated this revision most forcefully. One of the discoverers of the *hox* genes (responsible for laying out body plan), Gehring went on to make a second monumental discovery in 1995, in one of the most startling experiments in biology. Gehring's team took a gene from a mouse and inserted it into the fruit fly *Drosophila*. This was no ordinary gene, with a minor ensemble role: under its malign guidance the fruit fly suddenly started sprouting whole eyes on its legs, wings and even antennae (see Fig. 7.6). These strange diminutive eyes peeping out from peculiar places were not the familiar camera eyes of mice and men, but compound eyes, displaying all the arrays of facets characteristic of insects and crustaceans. What this gruesome experiment proved with visceral force was that the genes needed for growing an eye in a mouse and in a fly were the same: they had been preserved, with amazing fidelity, down 600 million years of evolution, ever since the last common ancestor of vertebrates and invertebrates, to the point that they were still interchangeable. Put the mouse gene in a fly and it took over the fly systems, wherever it was placed, commanding the subordinate hierarchy of fly genes to build an eye right there on the spot.

Nietzsche had once taught in Basel, and perhaps in homage, Gehring

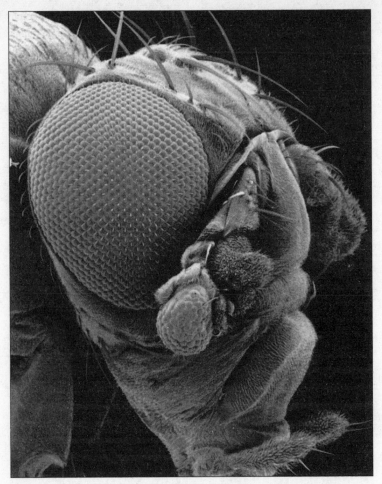

Figure 7.6 Scanning electron micrograph of the head of a fruit fly (*Drosophila*), showing a diminutive extra eye on the antenna, induced by genetically engineering a mouse *Pax6* gene. The same gene controls eye development in both vertebrates and invertebrates, and must have done so in their common ancestor, perhaps 600 million years ago.

referred to the mouse gene as a 'master gene'. I wonder whether 'maestro gene' might have been more appropriate; certainly less bombastic and perhaps more plural. Like an orchestral conductor conjuring up the most beautiful music without sounding a note himself, the gene calls forth the structures of the eye by ushering in individual players, each with their own part to play. Different versions of the same gene were already known through their mutations in flies, mice and men. In mice and flies, the gene was called *Small eye* and *Eyeless*, respectively, referring, in the inverted terms to which geneticists are horribly prone, to the degree of deficit in its absence. In our own case, mutations in the same gene cause the disease aniridia, in which the iris fails to develop; though an unpleasant and frequently blinding condition, a curiously limited outcome for a master gene supposed to supervise the whole construction of the eye. But that's if only one copy of the gene is damaged. If both copies are damaged or lost, the entire head fails to develop.

The picture has grown more complex since Gehring's seminal experiment. His 'master gene' is now known as *Pax6*, and is both more powerful and less lonely in its elevation than had seemed. *Pax6* has since turned up in practically all vertebrates and invertebrates, including the shrimp; a closely related gene is even found in jellyfish. And it turns out that *Pax6* is not only behind the formation of eyes but also large parts of the brain; hence the lack of head development when both copies are missing. At the same time, *Pax6* is not alone. Other genes, too, can summon up whole eyes in *Drosophila*; in fact it seems to be a peculiarly easy thing to do. These genes are all plainly related to each other and are very ancient. Most of them are found in both invertebrates and vertebrates, albeit with slightly differing roles and contexts. Sadly, the beautiful music of life is called forth not by a conductor but by a small committee.

The bottom line is that the same committee of genes controls eye formation in both vertebrates and invertebrates. Unlike rhodopsin, there is no practical 'engineering' reason for the process to be controlled by the same genes; they are all faceless bureaucrats and might just as easily have been a different bunch of faceless bureaucrats. The fact that it's always the same bunch (unlike the lens proteins, for example) betrays the hand of history, the quirkiness of happenstance rather than the force of necessity. And this history suggests that the photoreceptor cell evolved just once, in a common ancestor of the verte-brates and invertebrates, under the control of a small committee of genes.

There's another reason to believe that the photoreceptor evolved just once – the direct testimony of a living fossil. The survivor is a tiny marine ragworm, *Platynereis*, a few millimetres long and covered in bristles. A denizen of muddy estuaries and favoured bait of anglers, one wonders how many know that its overall shape and morphology has barely changed since Cambrian times. A worm like this was the common ancestor of both the vertebrates and invertebrates. Like all vertebrates and many invertebrates, the ragworm is bilaterally symmetrical: it's the same on both sides, unlike a starfish. This symmetry makes all of us *bilaterians*, an insect as much as you or me. Cru-cially, the ragworm evolved before this design, pregnant with potential, exploded into all the marvellous incarnations we see about us today. It is a living fossil of the primordial bilaterian, the ur-bilaterian, and that's why Detlev Arendt and his colleagues at the European Molecular Biology Lab in Heidelberg were interested in its photoreceptor cells.

They knew that the ragworm's eyes are similar in their design to the inver-tebrates, rather than the vertebrates, right down to the type of rhodopsin they use. But in 2004 the Heidelberg team discovered another clutch of light recep-tors, buried away in its brain. These were not used for seeing at all, but rather for the circadian clock, those internal rhythms that govern sleep and wakeful-ness and distinguish night from day, even in bacteria. Not only did the circa-dian clock cells use rhodopsin, but they were instantly recognisable (to experts like Arendt, at least) as *vertebrate* photocells, a recognition later confirmed by more detailed biochemical and genetic tests. The ur-bilaterian, Arendt con-cluded, possessed both types of photocell. And that meant that the two types hadn't evolved independently in totally different lines, but rather were 'sister' cells that had evolved together in the same organism – an ancestor of the ur-bilaterian.

Of course, if this common ancestor of vertebrates and invertebrates pos-sessed both types of photocell, then we might have inherited both, if only we knew where to look. And so, it seems, we did. The year after the living fossil offered up its secrets, Satchin Panda and his colleagues at the Salk Institute, San Diego, followed up a hunch about some cells in our own eye – the retinal ganglion cells – which influence human circadian rhythms. Although not specialised for light detection, they too possess rhodopsin. It's an unusual form known as melanopsin; and it turns out to be characteristic of *invertebrate* photocells. Remarkably, this circadian rhodopsin in our own eyes is closer in its structure to the rhodopsin in the naked retinas of the vent shrimp than it is to the other type of rhodopsin that shares the human retina.

All this implies that the vertebrate and invertebrate photocells sprang from the same source. They are not separate inventions, but sister cells with a mother in common. And that mother cell, the primordial photocell, the ancestor of all animal eyes, evolved just once.

The bigger picture that emerges, then, is this. A single type of light-sensitive cell, containing the visual pigment rhodopsin, evolved in a common ancestor of the vertebrates and invertebrates, under the control of a small committee of genes. Later, this light-sensitive cell was duplicated, and the two daughter cells became specialised to function either in eyes or in a circadian clock. For reasons that may have been no more than chance, the vertebrates and the invertebrates each selected opposing cell types for these tasks, so that eyes developed from different tissues in the two lines, giving rise to major embryonic differences between similar eyes in, for example, the octopus and mankind. The first station en route to the complete eye was a naked retina: a sheet of light-sensitive cells, composed of one or the other type of light-sensitive cell, according to the lineage. Some organisms still retain simple, flat, naked retinas, while in others the sheet became recessed into pits, able to cast a shadow and give a sense of where the light came from. As these pits deepened, the trade-off between sensitivity and resolution meant that any form of lens was better than none at all; and all kinds of unexpected materials, from minerals to enzymes, were recruited to the task. A similar process took place in different lines, giving rise to a cacophony of lens types; but the optical constraints on building a functional eye restricted this variety at the molecular level to a small range of large-scale structures, from camera eyes like our own to the compound eyes of insects.

There are, of course, innumerable details to fill in; but this, in broad brush stroke, is how the eye evolved. No wonder that we share the same rhodopsin with vent shrimp: we all inherited it from the same ancient ancestor. But that still leaves us with one big question to conclude this chapter – who was this ancestor? The answer, once more, lies in the genes.

Down in the deep-sea vents, Cindy Van Dover was worried about light. Her vent shrimp could apparently detect green light with extraordinary sensitivity, using a rhodopsin similar to that in our own eye; and yet the early measurements showed that the vents did not glow green. What was going on?

In a wry bit of advice to young scientists, offered during his retirement speech, an eminent researcher remarked that on no account should one ever repeat a successful experiment: it will certainly turn out a bitter disappointment.[8] The converse – never hesitate to repeat a failure – is less obviously true, yet Van Dover had good reason to try. Like a dead man, rhodopsin doesn't lie. If it absorbs green light, she reasoned, then there must be some green light there to absorb. Presumably, the rudimentary equipment used in the early studies was just not as sensitive as a shrimp's naked retina.

A new and altogether more sophisticated photometer was commissioned from the space scientists at NASA, who knew all about detecting radiation in the inky blackness of outer space. Named ALISS (Ambient Light Imaging and Spectral System), the device duly did detect light at other wavelengths. Down in the wonderland of the vent world, ALISS charted a small peak in the green part of the spectrum with an intensity orders of magnitude greater than predicted on the basis of theory. The new measurements were soon corroborated at other vents. Although the source of this eerie green glow is still a mystery, there is no shortage of exotic hypotheses. Small bubbles of gas emerging from the vents and crushed in the high pressure of the ocean can give rise to visible light, for instance, as can the formation and shock fracture of crystals under heat and pressure.

If Van Dover's faith in rhodopsin was well placed, she was only playing the odds. Rhodopsin has an impressive ability to track the conditions. The deep blue sea is so called because blue light penetrates further through water than do other wavelengths. Red light is soon absorbed by water, and can't penetrate far; yellow light gets a little further; orange further still. But from about twenty metres down, most light is green and blue, becoming bluer with depth. The blue light scatters around, making everything in the ocean deep a shade of blue. The rhodopsin pigments of fish track this blue shift with finesse, a trick known as spectral tuning. So around 80 metres down, we find fish that have rhodopsins which absorb green light best (at around 520 nanometres), but by 200 metres down, in the last fading embers of light, the fish possess rhodopsins that absorb blue light (at about 450 nanometres). Interestingly, the vent crab *B. thermydron*, which we met earlier, reverses this shift as it moves to the vent. The larvae of the crab live in deep blue waters and have a rhodopsin that absorbs blue light best, at a wavelength of 450 nanometres. In contrast, the naked retina of the adult has a rhodopsin that absorbs light at a wavelength of 490 nanometres, closer to the green. The shift is small but deliberate. Given that the rhodopsin of vent shrimp, too, absorbs green light at 500 nanometres, Van Dover's antennae had every reason to twitch.

Our own colour vision depends on the ability of rhodopsin to shift wavelengths. We have two types of photoreceptor in our retina, the rods and cones. Strictly speaking, only the rod cells contain rhodopsin; the cones contain one of three 'cone opsins'. But in reality this distinction is unhelpful, as all of these visual pigments are basically the same in structure: all are composed of a particular type of protein – an 'opsin', plugged across the membrane with a sevenfold zigzag – bound to a derivative of vitamin A called retinal. Retinal is a pigment, and as such is the only bit responsible for absorbing light. When it absorbs a photon it changes shape from a kinked to a straight form, and this is enough for it to set in motion the whole biochemical cascade that in the end signals *light!* to the brain.

Although it is the retinal that absorbs light, by far the most important factor for 'spectral tuning' is the structure of the opsin protein. Small changes in structure can shift its absorption from the ultraviolet (about 350 nanometres) in insects and birds to the red (about 625 nanometres) in chameleons. So by combining several slightly dissimilar opsins, each one with a different absorption, colour vision is possible. Our own cone opsins absorb light maximally in the blue (433 nanometres) green (535 nanometres) and red (564 nanometres) parts of the spectrum, together giving our familiar visible range.[9]

While the opsins are broadly similar in their overall structure, the differences between them divulge a fascinating history of life. All were formed by duplication followed by divergence, and can ultimately be traced back to an ancestral opsin gene. Plainly, some of these duplications happened more recently than others. Our 'red' and 'green' opsins are closely related, for example: the gene was duplicated in a common ancestor of the primates. This duplication gave the primates three types of cone opsin (or it did after they'd diverged a bit, anyway) rather than two, giving most of us three-colour (trichromatic) vision. A few unfortunates who are red–green colour-blind have lost one of these genes again, making them dichromatic, like almost all other mammals, their poor vision reflecting, perhaps, a relatively recent nocturnal past spent hiding from dinosaurs. Why the primates regained three-colour vision is disputed. The most popular theory suggests that it helped to spot red fruit against green leaves; an alternative, more socially oriented idea, argues it helped discriminate emotions, threats and sexual signals, from blushing to barefaced lies (and it's interesting that all trichromatic primates are bare-faced).

I said the primates 'regained' three-colour vision, but in fact we're still the poor relations among other vertebrates. Reptiles, birds, amphibians and sharks all have four-colour vision, and it seems likely that the common ancestor of the vertebrates was tetrachromatic, with an ability to see into the ultraviolet.[10] A lovely experiment confirmed the possibility: by comparing the gene sequences of living vertebrates, Yongsheng Shi and Shozo Yokoyama, at Syracuse University, New York, predicted the sequence of the ancestral vertebrate gene. As yet we have no way of guessing, from first principles alone, exactly what wavelength this ancestral rhodopsin would have absorbed. Nothing daunted, Shi and Yokoyama used genetic engineering techniques to build the protein, and then measured its light absorption directly. It duly turned out to be in the ultraviolet (360 nanometres).

The deepest branch in the tree of opsins lies between the vertebrates and the invertebrates, as we've seen. But even that living fossil, the ur-bilaterian ragworm *Platynereis*, still has two types of opsin, corresponding to those of the vertebrates and the invertebrates. So what did the grand ancestor of all animal opsins look like, and where did it come from? The answer is not known for sure, and several hypotheses jostle for prominence. But our guide has been the gene itself, and in using it we have traced our way back through 600

million years. How much further back can we go? According to Peter Hegemann and his colleagues at Regensburg University in Germany, the gene does
indeed give an answer, and one that is utterly unexpected: the earliest progenitor of the eye, they say, evolved in *algae*.

Algae, like plants, are masters of photosynthesis, and can call upon all
kinds of sophisticated light-sensitive pigments. Many algae use these pigments in simple eyespots to register the light intensity and, if necessary, to do
something about it. So, for example, the luminously beautiful algae *Volvox*
forms into hollow spheres composed of hundreds of cells, up to one millimetre across. Each cell wields two flagella that poke out like oars; these beat
in the dark, but stop in bright light, steering the whole sphere towards the sun,
tracking the best conditions for photosynthesis. The command to stop is controlled by the eyespots. The surprise is that the light-sensitive pigment in the
eyespots of *Volvox* is rhodopsin.

Even more unexpectedly, the *Volvox* rhodopsin looks as if it is ancestral to
all animal opsins. The site where retinal binds to the protein contains sections
that are exactly the same as both the vertebrate and the invertebrate opsins,
practically a mixture of each, in fact. And the overall structure of the gene,
with its eclectic mixture of coding and non-coding sequences (technically
known as introns and exons), also betrays an ancient link to both vertebrate
and invertebrate opsins. It's not proof, but it is exactly what one would predict
for an ancestor of both families. And that means there's a good chance that
the mother of all animal eyes was, of all things, a photosynthetic alga.

That, of course, raises the question: how on earth did algal rhodopsin get
into animals? Certainly the lovely *Volvox* is not on a direct line to the animals.
But a quick look at the eyespot structure immediately suggests a clue: the
rhodopsin is embedded in the membranes of the *chloroplasts*, those tiny structures in algae and plant cells responsible for photosynthesis. A billion years
ago, the ancestors of the chloroplasts were free-living photosynthetic bacteria, namely cyanobacteria, which were engulfed by a larger cell (see Chapter
3). That means that eyespots are not unique to *Volvox*, necessarily, but to the
chloroplasts, or perhaps even their forebears, the cyanobacteria.[11] And chloroplasts are found in many other types of cell too, including a few protozoa,
some of which *are* among the direct ancestors of the animals.

The protozoa are single-celled organisms, the best known of which is the amoeba. The seventeenth-century Dutch pioneer of the microscope Antony van Leeuwenhoek first saw them, along with his own sperm, and memorably referred to them as 'animalcules', distinguishing them from the microscopic algae, which he classified with plants as being basically vegetable. But this simple division hid a multitude of sins, for if we magnified some of these little animalcules up to our own size, we'd be terrorised by monsters that are half beast, half vegetable, staring back at us like the paintings of Arcimboldo. In more sober terms, some motile protozoa that swim around in pursuit of prey also contain chloroplasts, giving them an algal dimension, and indeed they acquired them in exactly the same way as the algae, by engulfing other cells. Sometimes these chloroplasts remain functional, backing up the dietary needs of their host; but in other cases they degenerate, leaving behind their characteristic membranes and genes as a fading memory of a once-glorious past, or, like the miscellaneous bits and pieces in the workshop of a tinker, the basis of a new invention, an invention, perhaps, like an eye. And exactly such a microscopic chimera, rather than *Volvox* itself, is the kind of creature that some researchers (notably Walter Gehring again) speculate may conceal the mother of all animal eyes.

Which tiny chimera? No one knows, but there are beguiling clues, and much to learn. Some protozoa (dinoflagellates) have astonishingly complex mini-eyes, with a retina, lens and cornea all packed into the same cell. These eyes seem to have developed from degenerate chloroplasts, and they too use rhodopsin. Whether animal eyes developed from them directly or indirectly (via a symbiosis) in this teeming and little known microcosm, is an open question. Whether it happened as a predictable step or as an outrageous freak of fortune, we can't say. Yet this kind of question, at once specific and universal, is the very stuff of science, and I hope that it will inspire a rising generation with eyes in their stars.

8

HOT BLOOD

Breaking the Energy Barrier

Time flies by when you're the driver of a train, runs a children's lyric. And who can't remember the reverse as a child – the endless minutes of mind-numbing tedium in the back of a car, asking repeatedly, 'Are we there yet, Daddy?' I imagine most readers will also remember the distress of watching their ageing grandparents, or parents, slow down to a snail's pace, in the end sitting inscrutably as hours pass by like minutes. Both extremes are far removed from the tempo of our own world, the andante of an adult human being.

We don't need Einstein to tell us that time is relative. But what Einstein established rigorously for time and space is, as ever, more impressionistic in biology. As the celebrated wag Clement Freud had it: 'If you resolve to give up smoking, drinking and loving, you don't actually live longer, it just seems longer.'[1] Yet there is a real sense in which time rushes through childhood, and crawls through old age. It lies in our internal settings, our metabolic rate, the rate at which our hearts beat and our cells burn up food in oxygen. And even among adults there are striking differences between the active and the slovenly. Most of us shift slowly from one to the other. The rate at which we slow down, or indeed gain weight, depends much on our metabolic rate, which varies innately between individuals. Two people who eat the same and exercise equally will often differ in their tendency to burn off calories while at rest.

Nowhere is metabolic rate more significant than the difference between hot-blooded and cold-blooded creatures. While these terms make biologists cringe, they are vivid and meaningful to almost everyone, and convey as much as the slippery technical terms, like homeothermy and poikilothermy. It's a curious thing, but I've noticed there are few aspects of biology that we feel so chauvinistic about, we hot-bloods. The fury and spleen vented in journals, and online, about whether dinosaurs, for example, were hot-blooded or cold-blooded is hard to understand rationally: it is a visceral distinction, perhaps something to do with our dignity, whether we would rather be eaten by giant lizards, or clever, scheming, fast-moving beasts, against whom we must pit our wits to survive. We mammals still bear a grudge, it seems, for the time we spent as small furry animals, cowering underground in hock to the top predators of the past. But then it was for 120 million years, which is a long time by any reckoning.

Hot blood is all about metabolic rate, all about the pace of life. Hot blood helps in its own right, for all chemical reactions speed up with rising temperature, including the biochemical reactions that underpin life. Over the small range of biologically meaningful temperatures, from around 0°C up to 40°C in animals, the difference in performance is striking. Oxygen consumption, for example, doubles with every 10°C rise in temperature in this range, corresponding to mounting stamina and power. So an animal at 37°C has twice the power of one at 27°C, and quadruple the power of an animal at 17°C.

But to a large extent, temperature misses the point. Hot-blooded animals are not necessarily any hotter than cold-blooded animals, for most reptiles are adept at absorbing the energy of the sun, warming their core body temperature up to levels similar to mammals and birds. Certainly, they don't maintain such high temperatures after dark; but then mammals and birds are often inactive at night too. They might as well save energy by lowering their core body temperature, but rarely do, at least not by much (although hummingbirds often pass into a coma to conserve energy). In our energy-conscious times, mammals ought to make environmentalists weep: our thermostat is jammed at 37°C, twenty-four hours a day, seven days a week, regardless of need. And forget alternative energy. We're in no way solar-powered, like lizards, but generate heat prodigiously by way of internal carbon-burning power stations, giving us a giant carbon footprint too. Mammals are the original eco-hooligans.

You might think that running on full power through the night would give mammals a head start in the morning, but lizards don't waste much time raising their temperatures back to operational levels. The earless lizard, for example, has a blood sinus on top of its head, through which it can warm its whole body rapidly. In the morning, it pokes its head out of its burrow, keeping a wary eye out for predators, ready to duck back in if necessary, and after half an hour is usually warm enough to venture out. It's a pleasant way to start the day. Characteristically, natural selection is not content with only one function. If caught out, some lizards have a connection from the sinus to their eyelids, through which they can squirt blood at predators, such as dogs, which find the taste repugnant.

Size is another way to maintain high temperatures. You don't need to be a great white hunter to picture the hides of two animals stretched out as rugs on the floor. Imagine that one such hide is twice the length and breadth of the other. This means that the larger animal had four times more hide than the smaller beast ($2 \times 2 = 4$), but it would have been eight times heavier, as it also had twice the depth ($2 \times 2 \times 2 = 8$). Thus every doubling of dimensions halves the surface-to-weight ratio ($4 \div 8 = 0.5$). Assuming that each pound in weight generates the same amount of heat, larger animals have more pounds and so generate more internal heat.[2] At the same time, they lose heat more slowly because their skin surface is relatively small (in relation to internal heat generated). So, the bigger the animal, the hotter it gets. At some point, cold-blooded creatures become hot-blooded. Large alligators, for example, are technically cold-blooded, but retain heat long enough to be borderline hot-blooded. Even overnight, their core temperature only drops a few degrees, despite producing little internal heat.

Plainly many dinosaurs would have surpassed this size threshold comfortably, making them de facto hot-blooded, especially given the pleasantly warm ambient temperatures enjoyed by much of the planet in those halcyon days. There were no ice caps, then, for example, and atmospheric carbon dioxide levels were as much as tenfold higher than today. In other words, some simple physical principles mean that many dinosaurs would have been hot-blooded, regardless of their metabolic status. The giant herbivores may well have had more trouble losing heat than gaining it; and some anatomical curiosities, like the great armoured plates of the stegosaurus, may have played a second role in heat dissemination, not unlike an elephant's ears.

But if it were as simple as that, there would have been no controversy about whether or not the dinosaurs were hot-blooded. In this limited sense they certainly were, or at least many of them were. For those who like mouth-filling terms, it's called 'inertial endothermy'. Not only did they maintain a high internal temperature, they generated heat internally, in the same way as modern mammals, through burning carbon. So in what broader sense were dinosaurs *not* hot-blooded? Well, some of them may well have been, as we'll see later, but to understand the real oddity of mammalian or avian hot blood we need to reverse the size trend to see what happens in smaller animals, below the 'hot-blood threshold'.

Think of a lizard. By definition, it is cold-blooded, which is to say, it can't maintain its internal body temperature overnight. While a large crocodile might come close, the smaller the animal, the harder it gets. Insulation, like fur or feathers, only helps to a point and can actually interfere with heat absorption from the surroundings. Dress up a lizard in a fur coat (and needless to say, earnest researchers have done exactly this) and the lizard gets steadily colder, unable to absorb the sun's heat so well, or to generate enough heat internally to compensate. This is far from the case with mammals or birds, and that brings us to the real definition of hot blood.

Mammals and birds generate up to ten or fifteen times as much internal heat as a similarly sized lizard. They do so regardless of circumstances. Place a lizard and a mammal in suffocating heat and the mammal will continue to generate ten times as much internal heat, to its own detriment. It will have to go out of its way to cool down – drink water, plunge into a bath, pant, find shade, fan itself, drink cocktails, or switch on the air-conditioning. The lizard will just enjoy it. It's not surprising that lizards, and reptiles in general, fare much better in the desert.

Now try placing the lizard and the mammal in cold conditions, let's say close to freezing, and the lizard will bury itself in leaves, curl up and go to sleep. To be fair, many small mammals would do that too, but that's not our default setting. Quite the contrary. Under such conditions, we just burn up even more food. The cost of living for a mammal in the cold is a hundred times that of a lizard. Even in temperate conditions, say around 20°C, a pleasant spring day in much of Europe, the gap is huge, around thirtyfold. To support such a prodigious metabolic rate, the mammal must burn up thirty times more food than a reptile. It must eat as much in a single day, every single day, as a lizard eats in a whole month. Given that there's no such thing as a free lunch, that's a pretty serious cost.

So there it is: the cost of being a mammal or a bird starts at around ten times the cost of being a lizard and is often far higher. What do we get for our expensive lifestyle? The obvious answer is niche expansion. While hot blood may not pay in the desert, it enables nocturnal foraging, or an active existence over winter in temperate climates, both of which are denied to lizards. Another advantage is brainpower, although it's hard to see why there should be a necessary relationship. Mammals certainly have far larger brains, relative to their body size, than reptiles. While a large brain is no guarantee of intelligence, or even quick wits, it does seem to be the case that a faster metabolism supports a bigger brain, without specifically dedicating resources to it. So if lizards and mammals both earmark, say, 3 per cent of their resources to the brain, but mammals have at their disposal ten times the resources, they can afford ten times more brain, and usually have exactly that. Having said that, primates, and especially humans, allocate a far greater proportion of their resources to brainpower. Humans, for example, dedicate around 20 per cent of resources to the brain, even though it takes up only a few per cent of our body. I suspect, then, that brainpower is little more than an added extra, thrown in at no extra cost, for a hot-blooded lifestyle. There are far cheaper ways of building bigger brains.

In short, niche expansion, nocturnal activity and added brainpower don't seem much payback for the serious metabolic costs of hot blood. Something seems to be missing. On the debit side, the costs of eating, eating, eating go well beyond bellyache. There is the serious cost of time and effort spent foraging, hunting or cropping vegetation, time vulnerable to predators or competitors. Food runs out, or becomes scarce. Plainly, the faster you eat, the faster you will run out of food. Your population shrinks. As a rule of thumb metabolic rate governs population size, and reptiles often outnumber mammals by ten to one. By the same token, mammals have fewer offspring (though they can dedicate more resources to the few they have). Even lifespan varies with metabolic rate. Clement Freud was right about people but wrong about reptiles. They may live slow and boring lives, but they do live longer, in the case of giant tortoises for hundreds of years.

So hot blood exacts a cruel toll. It spells a short life, spent eating dangerously. It depresses the population size and the number of offspring, two factors that should be penalised ruthlessly by natural selection. In recompense we have the boon of staying up at night and hanging out in the cold. That seems a poor deal, especially if we go to sleep anyway. Yet in the great pantheon of life, we routinely give top billing to the mammals and birds. What exactly is it that we have but the reptiles don't? It had better be good.

The single most compelling answer is 'stamina'. Lizards can match mammals easily for speed or muscle power, and indeed over short distances outpace them; but they exhaust very quickly. Grab at a lizard and it will disappear in a flash, streaking to the nearest cover as fast as the eyes can see. But then it rests, often for hours, recuperating painfully slowly from the exertion. The problem is that reptiles ain't built for comfort – they're built for speed.[3] As in the case of human sprinters, they rely on anaerobic respiration, which is to say, they don't bother to breathe, but can't keep it up for long. They generate energy (as ATP) extremely fast, but using processes that soon clog them up with lactic acid, crippling them with cramps.

The difference is written into the structure of muscle. There are various types of muscle, as we saw in Chapter 6. These vary in the balance of three key components: muscle fibres, capillaries and mitochondria. In essence, the muscle fibres contract to generate force, the blood capillaries supply oxygen and remove waste, while the mitochondria burn up food with oxygen to provide the energy needed for contraction. The trouble is that all of them take up valuable space, so the more muscle fibres you pack in, the less space there is left over for capillaries or mitochondria. A muscle packed tightly with fibres will have tremendous force, but soon runs out of the energy needed to fuel its contraction. It's a choice with the most widespread consequences – high power and low stamina, or low power and high stamina. Compare a bulky sprinter with a lean distance runner, and you'll see the difference.

We all have a mixture of muscle types, and this mix varies according to circumstances: whether we live at sea level or altitude, for example. Lifestyle can also make a big difference. Train to be a sprinter and you will develop bulky 'fast-twitch' muscles, with lots of power but little stamina. Train to be a long-distance runner and you'll shift the other way. Because these differences also vary innately between individuals and races, they are subject to selection over generations, if the circumstances dictate. That's why the Nepalese, East Africans and Andean Indians have a good many traits in common – traits that lend themselves to life at high altitude – whereas lowlanders are heavier and bulkier.

According to a classic paper in 1979 by Albert Bennett and John Ruben, then at the University of California, Irvine, such differences lie at the root of hot blood. Forget temperature, they said: the difference between hot-blooded and cold-blooded creatures is all about stamina. Their idea is known as the 'aerobic capacity' hypothesis, and even if it's not entirely right, it changed the way the whole field thought about life.

The aerobic capacity hypothesis makes two claims. First, selection is not for temperature but for increased activity, which is directly useful in many circumstances. As Bennett and Ruben put it themselves:

> The selective advantages of increased activity are not subtle but rather are central to survival and reproduction. An animal with greater stamina has an advantage that is readily comprehensible in selective terms. It can sustain greater levels of pursuit or flight in gathering food or avoiding becoming food. It will be superior in territorial defence or invasion. It will be more successful in courtship or mating.

That much seems incontestable. An interesting refinement of the idea, from the Polish zoologist Pawel Koteja, places the emphasis on intensive parental care, associated with feeding the young lasting for months or years, which sets mammals and birds apart from cold-blooded animals. Such investment requires very substantial stamina, and can have a big impact on survival at the most vulnerable time in an animal's life. Regardless of the exact reasons, though, it is the second part of the aerobic capacity hypothesis that is the more problematic and interesting: the link between stamina and rest. There is a necessary connection, say Bennett and Ruben, between the maximal and the resting metabolic rate. Let me explain.

The maximal metabolic rate is defined as the amount of oxygen consumed at full tilt, when we can push ourselves no further. It depends on many things, including fitness and, of course, genes. The maximal metabolic rate depends ultimately on the rate of oxygen consumption by the end-users, the mitochondria in the muscles. The faster they consume oxygen, the faster the maximal metabolic rate. But even a cursory reflection makes it plain that many factors must be involved, all of them interrelated. It will depend on the number of mitochondria, the number of capillaries supplying them, the blood pressure, the size and structure of the heart, the number of red blood cells, the precise molecular structure of the oxygen-transporting pigment (haemoglobin), the size and structure of the lungs, the diameter of the wind pipe, the strength of the diaphragm, and so on. If any one of these features is deficient, the maximal metabolic rate will be lower.

Selection for stamina, then, equates to selection for a high maximal metabolic rate, which boils down to selection for a whole suite of respiratory characters.[4] According to Bennett and Ruben, a high maximal metabolic rate somehow 'pulls up' the resting metabolic rate. In other words, an athletic mammal with lots of stamina has a high *resting* metabolic rate by default: it continues to breathe in plenty of oxygen, even while lying down doing nothing at all. They argued their case empirically. For whatever reason, they said, the maximal metabolic rate of all animals, whether mammal, bird or reptile, tends to be about ten times greater than the resting metabolic rate. Thus selection for high maximal metabolic rate drags up the resting metabolic rate too. If the maximal metabolic rate rises tenfold, which is the recorded difference between mammals and lizards, the resting metabolic rate also rises tenfold. And by that stage, the animal generates so much heat internally that it becomes, in effect, accidentally 'hot-blooded'.

The idea is pleasing and makes intuitive sense, but on closer examination it's very hard to work out quite why the two *need* to be linked. Maximal metabolic rate is all about getting oxygen out to the muscles, but at rest muscles contribute little to oxygen consumption. Instead, the brain and visceral organs – the liver, pancreas, kidneys, intestines, and so on – play the most important role. Exactly why the liver needs to consume lots of oxygen, just because the muscles do, is not clear. It is at least possible to imagine an animal that has a very high aerobic capacity and a very low resting metabolic rate, a kind of souped-up lizard that combines the best of both worlds. And it may be that this is exactly what the dinosaurs were. It's a bit of an embarrassment, frankly, that we still don't know why the maximal and resting metabolic rate tend to be linked in modern mammals, reptiles and birds, or if the link can be broken in some animals.[5] Certainly, very athletic mammals, like the pronghorn antelope, have very high aerobic capacities, around sixty-five times higher than their resting metabolic rates, implying that the two can be disconnected. The same applies to a few reptiles. The American alligator, for example, has an aerobic capacity at least forty times higher than its resting rate.

Be that as it may, there are still some good reasons to think that Bennett and Ruben are right. Perhaps the strongest relates to the source of heat in most hot-blooded animals. There are many ways to generate heat directly, but most hot-blooded animals don't bother: their heat production is an indirect consequence of metabolism. Only small mammals that lose heat rapidly, like rats, generate heat directly. Rats (and the young of many other mammalian species) make use of a specialised tissue known as brown fat, which is chock full of hot mitochondria. The trick they use is simple enough. Normally, mitochondria generate an electrical current, composed of protons, across their membrane, and this is used to generate ATP, the energy currency of the cell (see Chapter 1). The whole mechanism requires an intact membrane that acts as an insulator. Any leak in the membrane short-circuits the proton current, dissipating its energy as heat. And that's exactly what happens in brown fat – protein pores are deliberately inserted into the membrane, rendering it leaky. Instead of ATP, these mitochondria generate heat instead.

So if heat is the primary objective, the solution is leaky mitochondria. If all the mitochondria are rendered utterly leaky, as in brown fat, all the energy in food is converted into heat directly. It's simple and quick, and doesn't take up a lot of space, because a small amount of tissue generates heat efficiently. But that's not what normally happens. There's little difference in the degree of mitochondrial leakiness between lizards, mammals and birds. Instead, the difference between cold-blooded and hot-blooded creatures lies mostly in the size of the organs and the number of mitochondria. For example, the liver of a rat is much bigger than that of a similarly sized lizard and it's packed with far more mitochondria. In other words, the visceral organs of hot-blooded creatures are effectively turbocharged. They consume vast quantities of oxygen, not to generate heat directly, but to boost performance. Heat is merely a by-product, only later captured and put to good use with the development of external insulation, like fur and feathers.

The onset of hot-bloodedness in the development of animals today lends support to the idea that hot blood is more about turbocharging visceral organs than heat production. The evolutionary physiologist Frank Seebacher, at the University of Sydney, has begun looking into which genes underpin the onset of hot blood in embryonic birds, and finds that a single 'master gene' (which encodes a protein called $PGC1\alpha$) powers up the visceral organs by forcing their mitochondria to proliferate. Organ size, too, can be controlled quite easily, by adjusting the balance between cell replication and death, via similar 'master genes'. The long and short of it is that turbocharging the organs is not genetically difficult to do – it can be controlled by just a handful of genes – but it's energetically extremely costly, and will only be selected if the payback is worth it.

The broad scenario of the aerobic capacity hypothesis, then, looks convincing. There's no doubt that hot-blooded animals have far more stamina than cold-bloods, typically ten times the aerobic capacity. In both mammals and birds, this soaring aerobic capacity is coupled to a turbocharged resting metabolism – large visceral organs, with high mitochondrial power – but little deliberate attempt to generate heat. To me at least, it makes some sort of intuitive sense that a high aerobic capacity should be coupled to a boosted support system. And the idea is readily testable. Breed for high aerobic capacity, and the resting metabolic rate should follow suit. At the very least the two should correlate, even if causal relationships are hard to prove.

There's the rub. Since the hypothesis was proposed, nearly thirty years ago, there have been many attempts to verify it experimentally, with mixed success. There is indeed a general tendency for resting and maximal metabolic rates to be linked, but little more than that, and there are many exceptions to the rule. It may well be that the two *were* linked in evolution, even if such a link is not strictly necessary in physiological terms. Without a more specific idea of evolutionary history, it's hard to say for sure. But as it happens, this time the fossil record might actually hold the key. It may be that the missing link lies not in physiology, but in the vicissitudes of history.

Hot blood is all about the power of the visceral organs like the liver. Soft tissues don't survive the ravages of time well, though, and even fur is rarely preserved in the rocks. For a long time, then, it was hard to tease out the origins of hot blood in the fossil record, and even today angry controversy is rarely a stranger. But reappraising the fossil record in light of aerobic capacity is a more feasible task, since much can be gleaned from skeletal structure.

The ancestors of both mammals and birds can be traced back to the Triassic age, beginning 250 million years ago. The period came hot on the heels of the greatest mass extinction in the entire history of our planet, the Permian extinction, which is thought to have wiped out about 95 per cent of all species. Among the few survivors of that carnage were two groups of reptiles, the *therapsids* ('mammal-like reptiles'), ancestors of modern mammals, and the *archosaurs* (from the Greek 'ruling lizards'), the ancestors of birds and crocodilians, as well as dinosaurs and pterosaurs.

Given the later rise and dominance of the dinosaurs, it's perhaps surprising that the therapsids were the most successful group in the early Triassic. Their descendants, the mammals, shrank down in size and descended into holes before the onslaught of the dinosaurs. But earlier on in the Triassic, easily the most dominant species was *Lystrosaurus* ('shovel-lizard'), a pig-sized herbivore with two stumpy tusks, a squat face and a barrel chest. Quite what manner of life the lystrosaurs led is ambiguous. For many years they were pictured as amphibious beasts, a small reptilian hippo, but they are now thought to have lived in more arid climates, and purported to have burrowed holes, a common therapsid trait. We'll return to the significance of this later; but what is plain is that the lystrosaurs dominated the early Triassic in a way never seen again.[6] It's said that, for a period, 95 per cent of all terrestrial vertebrates were lystrosaurs. As the American poet and naturalist Christopher Cokinos put it: 'Imagine waking up tomorrow, walking across the continents and finding, say, only squirrels.'

The lystrosaurs themselves were herbivores, perhaps the only herbivores of that age, and feared no predators at the time. Later in the Triassic, a related group of therapsids called the *cynodonts* (meaning 'dog-teeth') began to displace the lystrosaurs, which finally fell extinct at the end of the Triassic 200 million years ago. The cynodonts included both herbivores and carnivores and were the direct forebears of the mammals, which emerged towards the end of the Triassic. The cynodonts showed many signs of high aerobic capacity, including a bony palate (separating the air-passages from the mouth, to allow simultaneous breathing and chewing), a broad chest with a modified rib cage and probably a muscular diaphragm. Not only that, but their nasal passages were enlarged, enclosing a delicate latticework of bone, known as 'respiratory turbinates'. The cynodonts might even have been covered in fur, but still laid eggs like reptiles.

It looks likely, then, that the cynodonts already had a high aerobic capacity, which must have given them great stamina; but what about their resting metabolic rate? Were they hot-blooded too? According to John Ruben, respiratory turbinates are one of the few reliable indications of an elevated resting metabolism. They restrict water loss, which can be very substantial during sustained heavy breathing, as opposed to short bursts of activity. Because reptiles have such a low resting metabolic rate, they breathe very gently when at rest and have little need to restrict water loss. As a result, no reptiles are known to have respiratory turbinates. In contrast, almost all true hot-blooded creatures do have turbinates, although there are a few exceptions including primates and some birds. Plainly turbinates help, even if they're not absolutely necessary, and their presence in fossils is as good a clue as any to the origin of hot blood. When coupled with the likely presence of fur (inferred rather than observed in fossils), it looks as if the cynodonts really did evolve hot blood somewhere along the line to the mammals.

But for all that, the cynodonts soon found themselves on the back foot, ultimately pressed into a cowering, nocturnal existence by the all-conquering archosaurs, in a late 'Triassic takeover'. If the cynodonts had evolved hot blood already, what about their vanquishers, a group that soon evolved into the first dinosaurs? The last survivors of the archosauran age, crocodiles and birds, are cold- and hot-blooded, respectively. At some point en route to the first birds, the archosaurs evolved hot blood. But which ones, and why? And did they include the dinosaurs?

Here the situation is more complex, and at times furiously controversial. Birds, like dinosaurs, attract passionate views that barely even masquerade as science. Long seen as related in some way to the dinosaurs, especially a group called the theropods that includes *Tyrannosaurus rex*, birds were redesignated squarely within the theropod line by a succession of systematic anatomical (cladistic) studies, dating back to the mid 1980s. The big conclusion was that birds are not merely related to the dinosaurs, they *are* dinosaurs, specifically avian theropods. While most experts are persuaded, a vociferous minority, led by distinguished paleo-ornithologist Alan Fedducia, at the University of North Carolina, maintain they derive from an earlier uncertain group that branched off before the evolution of theropods. In this view, birds are not dinosaurs; they are unique, a class unto themselves.

As I write, the latest in this long line of studies is also the most colourful, and relates to proteins rather than morphological traits. The amazing discovery, in 2007, by a team at Harvard Medical School led by John Asara, is that an exceptionally preserved bone from *T. rex*, some 68 million years old, still contains fragments of collagen, the main organic component of bone. The team succeeded in sequencing the amino acids in a few fragments, then piecing them together to give a sequence for part of the *T. rex* protein. In 2008, they compared this with equivalent sequences in mammals, birds and alligators. The sequences were short, and so potentially misleading, but on the face of it, the closest living relative of *T. rex* is the humble chicken, followed closely by the ostrich. Unsurprisingly, the reports were greeted by a chorus of approval in the newspapers, delighted to know finally how a *T. rex* steak would taste. More to the point, the collagen study broadly corroborates the cladistic picture of birds as theropod dinosaurs.

The other major source of rancour in the avian world is feathers. Feduccia and others have long maintained that feathers evolved for flight in birds, imparting to them a disturbingly miraculous sense of perfection. But if feathers evolved for flight, they certainly should not be found among non-avian theropods like *T. rex*. According to Feduccia, they're not; but a parade of feathered dinosaurs has marched out of China over the last decade. While some of these are a bit dubious, the majority of experts, again, are convinced that flightless theropods did indeed sport feathers, including a small ancestor of *T. rex* itself.

The alternative view, that the 'feathers' are not what they seem, but actually squashed collagen fibres, smacks of special pleading. If they were merely collagen fibres, it's hard to explain why they should be found mostly in a single group of theropods known as raptors, a group including *Velociraptor*, made famous by the film *Jurassic Park*. Or why they should look the same as the feathers of fully fledged birds, preserved in the same strata. Not only do the feathers look like feathers, but some raptors, notably *Microraptor*, could apparently glide between trees aided by feathers sprouting copiously from all four limbs (or, for want of a better word, wings). I find it hard to believe that these beautifully preserved feathers are not feathers; and even Feduccia is relenting. Whether the gliding arboreal flight of *Microraptor* has any bearing on the origins of flight in birds proper, or in their closest relative, *Archaeopteryx*, is a moot point.

The conclusion that feathers evolved in theropod dinosaurs, before the origin of flight, is backed by studies of the embryonic development of feathers in birds, and especially their relationship to the skin of embryonic crocodiles. Crocodiles, remember, are living archosaurs, the ruling lizards that first appeared back in Triassic times. The crocodiles and dinosaurs (including birds) started diverging in the mid-Triassic, around 230 million years ago. Yet despite this ancient divergence, crocs already held within them the 'seeds' of feathers; even today they retain exactly the same embryonic skin layers that develop into feathers in birds, as well as the selfsame proteins, called 'feather keratins', naturally light, flexible and strong.

The feather keratins are found mostly in some embryonic layers of crocodile skin that slough off after hatching to expose their scales (and remnants are found in the adult scales too). Birds have similar scales on their legs and feet, likewise exposed when the outer skin layers slough off after birth. According to Lorenzo Alibardi, a specialist in the evolutionary development of feathers at the University of Bologna, feathers grow from the same embry-onic layers that are sloughed off when scales form. The embryonic scales elongate into tubular filaments, or barbs. These are hollow hair-like struc-tures, with living walls formed from embryonic skin layers, which can sprout branches anywhere down their length.[7] The simplest feathers, down feathers, are basically tufts of barbs attached to the same spot, whereas flight feathers are formed from barbs that fuse into a central rachis. The living walls of barbs lay down keratin before degenerating to uncover a branching structure com-posed of keratin: a feather. Not only do the growing feathers co-opt existing skin layers and proteins, but even the genes needed are found in crocodiles, and so presumably were present in their shared archosaurian ancestors. Only the developmental programmes changed. The close embryological relation-ship between feathers and scales is betrayed by the (very) odd mutation that causes scales to erupt into feathers, which sprout from the legs of birds. Nobody has found a feathered crocodile yet, though.

From this perspective, prototype feathers are virtually bursting to get out of the skin of even the earliest archosaurs, so it's little surprise that theropods began sprouting 'epidermal appendages', probably ranging from bristles (like those of pterosaurs) to simple branching structures, akin to downy feathers. But what were they used for, if not flight? There are many plausible answers, by no means mutually exclusive, including sexual display, sensory functions, protection (barbs magnify size as well as potentially pricking like a porcupine) and, of course, insulation. The riot of feathered theropods certainly raises the possibility that they were hot-blooded, as their living relatives the birds.

Other evidence, too, squares with the idea of theropods as an active group of dinosaurs, at least implying that they had stamina. One feature is the heart. Unlike lizards and most other reptiles, crocodilians and birds all have power-ful hearts with four chambers. Presumably, then, the four-chambered heart was a trait inherited by all archosaurs, and therefore the dinosaurs too. A four-chambered heart is significant because it splits the circulatory system in two. Half supplies the lungs, the other half the rest of the body. This offers two important advantages. First, blood can be pumped at high pressure to the muscles, the brain, and so on, without damaging the delicate tissues of the lungs (leading to pulmonary oedema and probable death). Plainly, a higher blood pressure can support more activity, as well as far greater size. Large dinosaurs could never have pumped blood all the way up to their brains without the four-chambered heart. Second, splitting the circulatory system in half means there is no mixing of oxygenated with deoxygenated blood: oxygenated blood returns from the lungs and is immediately pumped at high pressure to the rest of the body, delivering maximal oxygen to the places of need. While the four-chambered heart doesn't necessarily imply hot blood (crocodiles are cold-blooded, after all), it's verging on the impossible to attain a high aerobic capacity without one.

The respiratory system of theropod dinosaurs also looks to have been similar to that of birds and could have supported high rates of activity. Bird lungs operate differently from our own, and are more efficient even at low altitudes. At high altitudes the difference is breathtaking. Birds can extract two or three times as much oxygen from rarefied air as mammals. That's why migrating geese can fly thousands of feet above the top of Everest, while mammals gasp for breath at much lower altitudes.

Our own lungs are built like a hollow tree, with air entering via the hollow trunk (the trachaea), and then following one of the branches (tracheoles) that lead into blind-ending twigs. The twigs don't end in sharp points, though, but in semi-inflatable balloons, the alveoli, which are riddled with tiny blood capillaries in their walls, the sites of gas exchange. Here the haemoglobin in the red blood cells gives up its carbon dioxide and picks up oxygen, before being spirited back off to the heart. The entire balloon system is inflated then deflated like bellows, through breathing, powered by muscles in the rib cage and diaphragm. The inescapable weakness is that the whole tree ends in dead spaces, where air barely mixes, just in the place fresh air is most needed. And even when fresh air does arrive, it has already mixed with the stale air on its way out.

Birds, in contrast, have a beautifully modified reptilian lung. The standard reptilian lung is a simple affair: just a big bag, really, divided by blade-like sheets of tissue, called septa, which partition the central cavity. Like the mammalian lung, the reptilian lung functions like bellows, either through expanding the rib cage, or, in the case of the crocodile, by way of a piston-like diaphragm attached to the liver and drawn back by muscles that fix on to the pubic bone. That makes the crocodile lung a bit like a syringe, where the diaphragm is equivalent to an airtight plunger that is drawn back to fill the lung. While this is quite a powerful method of breathing, the birds have gone even further and turned half of their bodies into a sophisticated one-way system of interconnecting air sacs. Rather than entering the lungs directly, air first flows into the air-sac system and eventually exits via the lungs, giving a continuous through-flow of air that eliminates the problem of dead space in our own blind-ending alveoli. Air flows past the septa (likewise refined in birds), during both inspiration and expiration, via the movement of lower ribs and the rear air-sac system – crucially, birds have no diaphragm. What's more, air flows one way, whereas blood flow is in the opposite direction, setting up a 'counter-current' exchange that maximises gas transfer (see Fig. 8.1).[8]

The question that has divided the field in acrimony for decades is, what kind of lungs did the theropods have? Piston lungs, like crocodiles, or through-flow lungs, like birds? The air-sac system in birds invades not only the soft tissues of the abdomen and chest but also the bones, including the ribs and spine. It has long been known that the theropods have hollows in their bones in the same places as birds. The incendiary palaeontologist Robert Bakker used this finding, among others, to reconstruct dinosaurs as active hot-blooded animals in the 1970s, a revolutionary view that inspired Michael Crichton's book, and later the film, *Jurassic Park*. John Ruben and colleagues, though, reconstructed theropod lungs differently, much closer to crocodiles, with a piston diaphragm arguably identifiable in one or two fossils. Ruben doesn't deny the existence of air pockets in theropod bones, just their purpose. They were not there to provide ventilation, he said, but for other reasons: to reduce weight or aid balance in bipedal animals, perhaps. The dispute grumbled on, incapable of proper resolution without new data, until the publication of a landmark paper in *Nature*, in 2005, by Patrick O'Connor and Leon Claessens, then at Ohio University and Harvard, respectively.

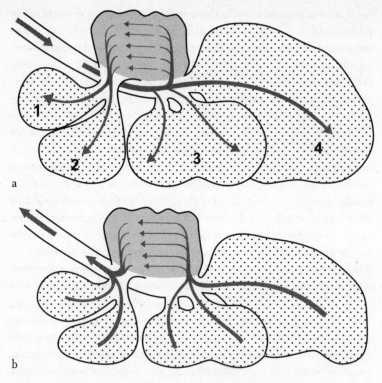

Figure 8.1 Air flow through bird lungs during (a) inspiration and (b) expiration. 1: clavicular air sac; 2: cranial thoracic air sac; 3: caudal thoracic air sac; 4: abdominal air sac. Air flows continuously in the same direction through the lung, while blood flows the other way, giving a highly efficient counter-current exchange of gases.

O'Connor and Claessens started out conducting a thorough examination of the air-sac systems of several hundred living birds (or rather, as they said, 'salvage specimens' taken from wildlife rehabilitators and museums). They injected the air sacs of these birds with latex, to get a better sense of pulmonary anatomy. Their first realisation was that the system is even more pervasive than they had appreciated, occupying not only parts of the neck and chest, but also much of the abdominal cavity, from where it invades the lower spine, a detail that was critical to interpreting the skeletal anatomy of theropods. This rear (caudal) air sac is the real driving force behind the whole pulmonary system of birds. During breathing, it becomes compressed, squeezing air into the lungs from behind. On expanding again, the caudal air sac sucks in air from the connecting air sacs in the chest and neck. In the lingo, it's an *aspiration pump*. It works a bit like bagpipes, in which pumping the bag streams a continuous airflow through the chanter.

O'Connor and Claessens went on to apply their findings to the bone structure of fossil theropods, including a superb skeleton of *Majungatholus atopus*, a theropod only distantly related to birds. While most studies have focused on the bone structure of the upper vertebrae and the ribs, they looked for hollows in the lower spine, as evidence of abdominal air sacs in theropods, and duly found them, in exactly the same place as in birds. Not only that, but the anatomy of the spine, rib cage and sternum met the specifications of an aspiration pump: the greater flexibility of the lower ribs and sternum allows compression of a caudal air sac, able to ventilate the lungs from behind as in birds. All in all, there can be little doubt that the theropod dinosaurs really did have an aspiration pump like that of birds – the most efficient system of breathing in all vertebrates (see Fig. 8.2).

So theropods had feathers, four-chambered hearts and air sacs coupled to through-flow lungs, all of which suggest they lived active lives, requiring stamina. But did their stamina lead inevitably to proper hot blood, as argued by the aerobic capacity hypothesis, or were they a halfway house, intermediate between modern crocodiles and birds? While their feathers suggest insulation, and so hot blood, they could have served other purposes instead; and further evidence, including the respiratory turbinates, is more ambiguous.[9]

Birds, like mammals, mostly possess respiratory turbinates, yet they are not composed of bone like the mammalian variety, but cartilage, which does not preserve well. Thus far, there's been no sign of turbinates in theropods, though few fossils are well enough preserved to judge. More tellingly, however, John Ruben notes that the turbinates in birds are invariably associated with enlarged nasal passages. Presumably, the delicate scrollwork of turbinates impedes airflow to a degree, which can be offset by enlarging the passages. But the theropods don't have especially large nasal passages, and

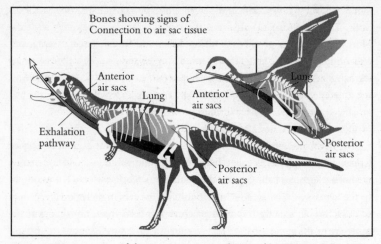

Figure 8.2 Reconstruction of the air-sac system in a dinosaur like *Majungatholus atopus* compared with modern birds. In both cases, the lung is supported by anterior and posterior air-sacs, the traces of which are exactly analogous to birds in dinosaur bones. The air-sacs work like bellows to move air through the rigid lung.

that implies that the apparent absence of turbinates is real, not merely a preservation artefact. If they didn't have turbinates, could they have been hot-blooded? Well, we don't have turbinates, and we are hot-blooded, so the answer technically is yes; but it does raise some questions.

Ruben himself believes that theropods *did* have high aerobic capacity, but not hot blood, despite his own aerobic capacity hypothesis stipulating that the two must be linked. And while we don't yet know enough to say for sure, the consensus position, insofar as there is one, is that theropods probably had a raised resting metabolism, but not yet true hot blood. That at least is the story of the fossils, but there is more in the rocks than fossils, including a record of ancient climates and atmospheres. And there was something about the air in the Triassic that places quite a different spin on the fossil record. It helps to explain not only the high aerobic capacity of cynodonts and theropods, but also why the dinosaurs sprang to dominance.

.✦.

Most discussions of physiology take place in a historical void: an unspoken assumption that the past was the same as the present, selection pressures as unchanging as gravity. But they are not, as the great extinctions attest. And the greatest of all extinctions came at the end of the Permian age, some 250 million years ago, the immediate curtain-raiser to the irresistible rise and rise of the ruling lizards and the ensuing age of the dinosaurs.

The Permian extinction is often regarded as one of life's great mysteries – apart from anything else, that helps attract grant monies – but the environmental background has been sketched out in broad brushstroke. It was not, in fact, one mass extinction, but two, separated by nearly 10 million years, an age of desperate decline. Both extinctions corresponded in time to prolonged volcanic unrest, the most extensive outpourings of lava in the history of the earth, burying vast areas, almost continental, beneath deep basalt. The lava flows eroded to form a stepped terrain known as 'traps'. The first of the volcanic episodes forged the Emeishan traps in China, around 260 million years ago, followed 8 million years later by a second, even greater outpouring, that produced the Siberian traps. Critically, both the Emeishan and Siberian volcanic flows erupted through strata containing carbonate rocks and coal. That's significant because the intensely hot lava reacts with carbon to release massive quantities of carbon dioxide and methane, each and every eruption for thousands of years.[10] And that changed the climate.

There have been lots of attempts to oust the killer behind the Permian extinctions, with strong cases made for global warming, ozone depletion, methane release, carbon dioxide suffocation, oxygen deprivation, hydrogen sulphide poisoning, and so on. The only case more or less ruled out is meteorite impact; there's little evidence of an impact like the one that finally brought the curtains down on the long reign of the dinosaurs, nearly 200 million years later. All the rest of the list, though, are more than plausible, and the big advance in the last few years has been the recognition that all of them are intimately and irrevocably linked. Any volcanic episode on the scale of the Emeishan traps sets in motion a train of circumstances that unfurls with inexorable momentum, a progression to chill the heart. Similar trains of interdependence threaten our own world today, though not, as yet, in a way that begins to compare.

The volcanoes belched methane and carbon dioxide high into the stratosphere, along with other noxious gases, which damaged the ozone layer and ultimately warmed and dried the world. The arid lands spread across the vast continent of Pangaea. The great coal swamps of the preceding periods, the Carboniferous and Permian, dried out, and began to blow in the winds, their carbon consumed by oxygen, drawing down the vitality of the very air. Over 10 million years, oxygen levels plunged, a crash in slow motion, from 30 per cent, to a trough below 15 per cent. The combination of warming waters (limiting the solubility of oxygen), falling atmospheric oxygen, and high carbon dioxide choked life from the seas. Only bacteria thrived, a poisonous sort that once dominated our planet in the age before plants and animals, spewing out the toxic gas hydrogen sulphide in oceanic quantities. The seas turned black and lifeless. Gas belching from the deadening oceans corrupted the air still further, suffocating animals on the shores. And then, only then, came the final hammer blows of fate, the eruption of the great Siberian traps, a death-knell striking once and again over 5 million years. For those 5 million years or more, little stirred in the seas or on land; and then began the first glimmers of recovery.

Who survived? The answer, curiously, is much the same at sea as on land: those that were the best at breathing, those that could cope with low oxygen, high carbon dioxide, and a nasty mix of noxious gases. Those that were equipped to gasp for breath and yet still remain active, those that lived in holes, in burrows, in slime, in bogs, in sediments, those that scavenged their living in places where nothing else wanted to be. A thousand thousand slimy things lived on, and so did we. And that's why it's significant that the first land animals to recover after that great dying were the lystrosaurs, those burrowers with barrel chests, muscular diaphragm, bony palate, widened air passages and respiratory turbinates. They emerged, panting, from their rancid burrows, and colonised the empty continents like squirrels.

This amazing story, written in the chemistry of the rocks, goes on for millions of years – it was the stamp of the Triassic age. The toxic gases disappeared, but carbon dioxide soared, ten times higher than today. Oxygen remained stubbornly below 15 per cent, the climate endlessly arid. Even at sea level, animals gasped for oxygen, a mouthful of air as thin there as at high altitudes today. This was the world of the first dinosaurs, hauling themselves on to their hind legs, freeing their lungs from the constraints of sprawling lizards that can't walk and breathe at once. Couple that with air sacs and an aspiration pump, and the rise of the dinosaurs begins to look inevitable, a story plotted out in convincing detail in an important book, *Out of Thin Air*, by palaeontologist Peter Ward, at the University of Washington. Archosaurs displaced cynodonts, says Ward (and I believe him), because the septate lung held within it the secret of success, an unknowable, latent ability to transform into the wonderful through-flow lungs of birds. The theropods were the only animals alive that didn't need to pant all the time. They had little need for turbinates.

And so stamina was no added extra, but a lifesaver, a ticket with the winning number to survive through terrible times. But this is where I part company with Ward, reluctantly. I agree that high aerobic capacity must have been critical to survival, but would it really have dragged up resting metabolic rate too? Ward implies this (by citing the aerobic capacity hypothesis), but that's not what happens today when animals live at high altitude. On the contrary, muscle mass tends to fall and wiry builds win. Aerobic capacity might be high, but resting metabolic rate does not rise in synchrony; if anything, it falls. Physiology in general is parsimonious in hard times, not profligate.

Back in the Triassic, with survival at a premium, did animals really raise their resting metabolic rate unnecessarily? That sounds counterintuitive at the least. The theropods seem to have raised their aerobic capacity without needing to become fully hot-blooded, at least at first. And yet the vanquished cynodonts apparently did become hot-blooded. Did they do it to compete, with little hope of success, against the formidable archosaurs? Or did it help them to remain active as they shrank down in size and took to the nights? Both are perfectly credible possibilities, but there's another answer I like even better, an answer that may shed some light on why the dinosaurs did precisely the reverse, burgeoning into giants the like of which the world has never seen again.

Vegetarians, in my experience, have a bad tendency to be holier than me; or perhaps it's just my carnivorous sense of guilt. But according to a quietly significant paper that slipped into a quietly obscure journal, *Ecology Letters*, in 2008, the vegetarians may have a lot more to be smug about than I've given them credit for. If it weren't for vegetarians, or rather their ancestral herbivores, we may never have evolved hot blood, and the fast pace of life that goes with it. The paper is by Marcel Klaassen and Bart Nolet at the Netherlands Institute of Ecology and it takes a splendidly numerate (technically 'stoichiometric') line on the difference between meat and greens.

Say the word 'protein' and most people think about a mouth-watering steak; and there is indeed a very strong connection in the mind, arising no doubt from our endless cookery shows and dieting manuals, between proteins and meat. Eat meat for proteins, and if you're a vegetarian make sure you eat plenty of nuts, seeds and pulses. Vegetarians, by and large, are more aware of dietary composition than meat-eaters. We need to eat proteins to ensure that we get enough nitrogen in our diet, which is needed for making fresh proteins for ourselves, as well as DNA, both rich in nitrogen. We actually have very little problem maintaining a balanced diet, even if we're vegetarian, but then we are hot-blooded: we eat a lot, by definition. Klaassen and Nolet point out that this is not at all true for cold-blooded animals. They don't eat a lot, by definition, and that gives them an interesting problem.

Very few contemporary lizards are herbivores, and of all the 2,700 species of snake not a single one is herbivorous. Of course, some lizards are herbivores, but they tend to be either relatively large, like iguanas, or given to greater activity, and higher body temperatures, than carnivorous lizards. Unlike the meat-eaters, which are quick to lower their body temperature and slump into a dormant state if need be, herbivorous lizards are far less flexible and have to soldier on. This has traditionally been ascribed to the difficulties of digesting plant materials, achieved with the aid of gut microbes able to ferment obstinate plant material, a process that works much better at higher temperatures. According to Klaassen and Nolet, though, there may be another reason, relating to the nitrogen content of typical plant matter. They performed an inventory of dietary nitrogen and confirmed that herbivorous lizards do indeed have a serious problem.

Imagine eating only greens, lacking in nitrogen. How can you get enough nitrogen in your diet? Well you could try to eat more widely, scavenge a little, eat seeds, and so on, but even then you will probably fall short. Or you could simply eat more. If you consume only, say, a fifth of your daily nitrogen needs by eating a bucket of leaves, then all you need to do is eat five buckets. If you do that you'll be left with a surplus of carbon, which plant matter is rich in, and you'll need to get rid of it somehow. How? Just burn it, say Klaassen and Nolet. A strictly herbivorous diet is perfectly attainable for hot-blooded animals, because we burn off bucket-loads of carbon all the time; but it's always problematic for cold-blooded animals. And in this context we might do well to look again at the lystrosaurs, which were herbivores, and the cynodonts, which were a mixture of herbivores and carnivores. Might it be that hot blood evolved in the cynodonts because they had a high aerobic capacity, a prerequisite of survival in those thin times, coupled with a diet rich in greens? Once hot blood had evolved in these early herbivores, they might easily have taken advantage of the extra energy to recover quickly, to roam for miles over the arid Triassic lands in search of food or in flight from predators. Predators had less of a dietary need for hot blood, perhaps, but they had to compete with those turbocharged herbivores on equal terms. Perhaps they needed hot blood just to keep up with the flight of the vegetarian Red Queen.

But what of the colossal dinosaurs, the most famous herbivores in history? Did they follow an alternative strategy to attain the same ends? If you eat five buckets of leaves but don't burn it off constantly, you could simply store it somewhere: get bigger, become a giant! Not only do giants have more 'storage capacity', they also invariably have a lower metabolic rate, which equates to a slower turnover of proteins and DNA, lowering the dietary need for nitrogen. So there are two plausible ways of coping with a diet rich in greens: larger size coupled with slower metabolism, or smaller size coupled with faster metabolism. It's revealing that these are exactly the strategies adopted by herbivorous lizards today, although they may be precluded from attaining a true hot-blooded state by their inherently low aerobic capacity. (How these lizards survived the Permian extinction is another question for another place.)

But why, then, did the dinosaurs get so big? The question has never been answered pleasingly despite many attempts. According to a throwaway line in a 2001 paper by Jared Diamond and his colleagues, the answer might conceivably lie in the high carbon dioxide levels at the time, which probably induced greater primary productivity, that is, faster plant growth. What Diamond's insight lacked, though, was the perspective on nitrogen provided by Klaassen and Nolet. High carbon dioxide levels do indeed induce greater productivity, but they also lower the nitrogen content of plant matter, a field of research that has grown up around concerns about the effect that rising carbon dioxide levels, in our own age, might have on feeding the planet. And so the problem facing the cynodonts and the dinosaurs was even more acute then than it is today: to get enough nitrogen in their diet, they needed even more greens. Strict vegetarians would have needed to eat gargantuan quantities.

And perhaps this explains why the theropods didn't need hot blood. They were carnivores, and so didn't face a nitrogen-balance problem. But unlike panting cynodonts, obliged to compete on equal terms with turbocharged herbivores, the theropods were above all that. They had super-efficient aspiration-pump lungs and they could catch anything that moved.

It wasn't until later, in the Cretaceous era, that the odd raptor turned to vegetarianism. And one of the first, as it happens, was a maniraptoran called *Falcarius utahensis*, described formally in *Nature* in 2005 by a team from Utah and informally by one of the authors, Lindsay Zanno, as 'the ultimate in bizarre, a cross between an ostrich, a gorilla and Edward Scissorhands'. But it was a *bona fide* missing link – half raptor, half herbivore – and lived around the time of the first tasty flowering plants, a time of unprecedented enticement to a vegetarian way of life. But from our point of view in this chapter, perhaps the most significant fact about *Falcarius* is that it was part of a group, the maniraptorans, from which birds are thought to have evolved. Could it be that the evolution of hot blood in birds, too, was linked with a shift in diet to vegetarianism, and so a greater dietary need for nitrogen? It's not totally implausible.

This chapter is closing on a speculative note. But speculation is easily dressed up as hypothesis, once described by Peter Medawar as an imaginative leap into the unknown; and that is the basis of all good science. There is much here that remains to be examined or tested, but if we want to unravel the reasons for our fast-paced way of life, we may need to look beyond the principles of physiology, and into the story of life itself – to a time in the history of our planet when extreme circumstances played a magnified role. Perhaps this is history, more than science, in that events didn't need to be so, they just happened that way. If the Permian extinction had never happened, or its prolonged low-oxygen aftermath, would high aerobic capacity ever have been a matter of life and death? Would life have bothered to go beyond the primitive reptilian lung? And if a few of these aerobically charged animals had not turned vegetarian, would hot blood exist? Perhaps this is history, but reading that remote past is a science in its own right, one that can only enrich our understanding of life.

CONSCIOUSNESS

Roots of the Human Mind

In 1996, Pope John Paul II wrote a celebrated message to the Pontifical Academy of Sciences, in which he recognised that evolution is more than a hypothesis. 'It is indeed remarkable that this theory has been progressively accepted by researchers, following a series of discoveries in various fields of knowledge. The convergence, neither sought nor fabricated, of the results of work that was conducted independently is in itself a significant argument in favour of this theory.'

Perhaps not surprisingly, though, the Pope was not about to throw the baby out with the bathwater. The human mind, he said, was forever beyond the domain of science. 'Theories of evolution which, in accordance with the philosophies inspiring them, consider the mind as emerging from the forces of living matter, or as a mere epiphenomenon of this matter, are incompatible with the truth about man. Nor are they able to ground the dignity of the person.' Inner experiences and self-awareness, he said, all the metaphysical apparatus through which we communicate with God, are impervious to the objective measurements of science, falling instead within the realms of philosophy and theology. In short, while conceding the reality of evolution, he was careful to discriminate the Magisterium of the Church as above evolution.[1]

This is not a book about religion, and I have no wish to attack anyone's devoutly held beliefs. Nonetheless, for exactly the same reasons that the Pope was writing about evolution ('The Church's Magisterium is directly concerned with the question of evolution, for it involves the conception of man') scientists are concerned with mind, for that involves the conception of evolution. If the mind is not a product of evolution, what actually is it? How does it interact with the brain? The brain is obviously physical, so presumably it is the product of evolution like animal brains, which share many, if not all, structures. But if so, does the mind evolve as the brain evolves, for example during the expansion of brain size in hominid skulls over the last few million years (certainly not a bone of scientific contention)? For that matter, how do matter and spirit interact at a molecular level, as they would have to; for otherwise how could drugs or brain injuries affect the mind?

Steven Jay Gould wrote positively of two non-overlapping Magisteria, Science and Religion, yet there are inevitably a few places where the twain must meet and overlap, consciousness being the prime example. These issues plumb the depths of history. Descartes, in proposing a split between spirit and substance, was in reality doing no more than formalising an idea with roots in antiquity and favoured by the Church – as a devout Catholic, he had no stomach for the condemnation meted out to Galileo by the Church. By formalising the split, Descartes freed the body, even the brain, for scientific study. Unlike the Pope, few scientists today are out-and-out Cartesian dualists, in the sense of believing in a separation between spirit and substance, but the concept is not ludicrous, and the questions I pose above are susceptible to scientific exploration. Quantum mechanics, for example, still holds open the door to deeper cosmic mysteries of mind, as we shall see.

I'm quoting the Pope because I think that what he says goes beyond religion, into the heart of man's conception of himself. Even those who are not religious may feel that their spirit is somehow immaterial, uniquely human, and in some way 'beyond science'. Few people who've read this far will feel that science has no right to pontificate on consciousness, and yet perhaps equally few would give evolutionists any special rights over a ruck of other disciplines that can claim insight – robotics, artificial intelligence, linguistics, neurology, pharmacology, quantum physics, philosophy, theology, meditation, Zen, literature, sociology, psychology, psychiatry, anthropology, ethology, and more.

I should say at the outset that this chapter is different from the other chapters in this book, in that not only does science not (yet) know the answer, but at present we can barely conceive of how that answer might look in terms of the known laws of physics or biology or information. There is no agreement among scholars of the mind about exactly how the firing of neurons could give rise to intense personal sensations.

But that is all the more reason to enquire what science can tell us about the workings of the human mind, and where those efforts meet a wall of unknowing. The Pope's position strikes me as defensible, insofar as we do not know how 'mere matter' generates the perceived immateriality of mind; indeed, we don't even know what mere matter actually is, or why matter exists, rather than nothing at all (in some ways a similar question to that of why consciousness exists, rather than non-conscious information processing). However, I think, or perhaps I should say I believe, that evolution does explain the most ethereal monuments of mind.[2] And more: the known workings of the human mind are so much more marvellous than the untutored mind can even begin to imagine that there is every reason to ground the dignity of the person in the majesty of the biological mind.

There are other compelling reasons for science to take up the challenge. The human mind is not always the rich vessel that we treasure. Diseases of the brain strip away the workings of the mind. Alzheimer's disease cruelly peels back the layers of a person, revealing ultimately their innermost lack of being. Deep depression is far too common, a malignant sadness that consumes the mind from within. Schizophrenia pulls the most real and harsh illusions, while some epileptic seizures dissolve the conscious mind altogether, exposing the zombie within. These conditions give a chilling impression of the vulnerability of the human mind. Francis Crick famously observed that 'you're nothing but a pack of neurons'; he might have added that they build a fragile house of cards. For society, for medicine, not to strive to understand and try to cure such conditions would be to deny the very charity that is esteemed so highly by the Church.

The first problem faced by any scientific account of consciousness is definition: consciousness means all things to all people. If we define consciousness as the awareness of *self* embedded in the world – a rich autobiographical awareness that defines an individual in the context of society and culture and history, with hopes and fears for the future, all cloaked in the dense, reflective symbolism of language – if this is consciousness, then of course mankind is unique. There is a chasm between humans and animals, none of which can be graced with the word, nor even our own ancestors or young children.

Perhaps the apotheosis of this view came in a strange book, *The Origin of Consciousness in the Breakdown of the Bicameral Mind*, by the American psychologist Julian Jaynes. He sums it up nicely: 'At one time, human nature was split in two, an executive part called a god, and a follower part called a man. Neither part was consciously aware.' What is surprising is how recently Jaynes places that period – some time between the composition of the *Iliad* and the *Odyssey*. (Of course, Jaynes takes these very different epics to be composed by different 'Homers', hundreds of years apart.) The essential point is that consciousness, for Jaynes, is purely a social and linguistic construct, and a recent one at that. The mind is conscious only when it becomes *aware* that it is conscious: when the penny drops. As an argument that's fine, but any argument that sets the bar so high as to exclude the author of the *Iliad* is surely too high. If Homer the elder was not conscious, was he then some unconscious zombie? If not, there must be a spectrum of consciousness, in which the highest form is self-awareness as a free and literate member of society, and lower forms are simply lower.

Most neuroscientists make a distinction between two forms of consciousness, which have their roots in the structure of the brain. The terms and definitions vary, but essentially 'extended consciousness' refers to the full glories of the human mind, utterly unattainable without language, society, and so on; while 'primary' or 'core' consciousness is something altogether more animalistic – emotions, motivations, pain, a rudimentary sense of self lacking an autobiographical perspective or a sense of death, and an awareness of objects in the world. The world of a fox that, when caught in a jawed trap, gnaws off its own leg to escape. As the distinguished Australian scientist Derek Denton observes in his fine book on animal consciousness, *The Primordial Emotions*, surely the animal is aware that it is held by the trap jaws and has an intention to get free. It has some awareness of self, and it has a plan.

The irony is that extended consciousness is relatively easy to explain, even if the word 'easy' should be qualified. Given a low-grade sense of 'aware-ness', there is nothing about extended consciousness that transgresses our physical understanding of the world; there is just a daunting parallel circuitry in the brain, embedded in the complex setting of society. There is nothing miraculous about society itself, for example. Plainly a child who is raised in isolation in a cave will possess no more than rudimentary consciousness, just as we may suspect that a Cro-Magnon child, raised in Paris today, would be indistinguishable from the French. Likewise with language. Most people find it impossible to conceive of any form of developed consciousness in a person or species lacking language, and again that is almost certainly true. But there's nothing magical about language. Language can be programmed into a robot sufficiently well to pass an intelligence test (such as the Turing test) without the robot ever becoming 'conscious' or possessing even a basic awareness. Memory, too, is eminently programmable; thank God my computer can remember every word I type. Even 'thinking' is programmable – just consider the chess-playing computer 'Deep Thought' (named from *The Hitchhiker's Guide to the Galaxy*) and its successor, 'Deep Blue', which in 1997 defeated the reigning world champion Gary Kasparov.[3] If humans can program these things, so too can natural selection, of that there can be no doubt.

I don't want to belittle the importance of society, memory, language and reflection to human consciousness: obviously it feeds on them all. The point is that, to be conscious, all of them depend on a deeper form of consciousness – feelings. It's easy to imagine robots with the brainpower of Deep Blue, with language, with sensors of the external world, with a near-infinite memory, but with no consciousness. No joy, no sorrow, no love or sadness of parting, no exultation of understanding, no hope, faith or charity, no thrill of a delicate scent or of lightly glancing flesh, no warmth of the sun on the back of your neck, no poignancy of the first Christmas away from home. Perhaps one day a robot will feel all this in its cogs, but for now we don't know how to program poignancy.

This is the same inner life ring-fenced by the Pope as falling within the Magisterium of the Church, and was famously described at about the same time by the Australian philosopher David Chalmers as the 'hard problem' of consciousness. Since then there have been many attempts to address the problems of consciousness, some quite successfully; but none has successfully addressed Chalmers' hard problem. Even the iconoclastic philosopher Daniel Dennett, accused of denying the problem altogether, actually sidesteps it in his celebrated 1991 opus, *Consciousness Explained*. Why shouldn't neurons firing feel of something, he asks at last, in closing his chapter on qualia (subjective sensations)? Why not indeed; but doesn't that just beg the question?

I am a biochemist, and I know its limitations. If you want to explore the role of language in fashioning consciousness, read Steven Pinker. I didn't include biochemistry in my list of subjects that can lay claim to any expertise in consciousness. Remarkably few biochemists have ever tried seriously to tackle consciousness, Christian de Duve being a possible exception. And yet surely Chalmers' hard problem is actually a problem in biochemistry. For how does the firing of neurons generate a 'feeling' of anything? How do calcium ions rushing through a membrane generate the sensation of red, or fear, or anger, or love? Let's keep this question in mind as we explore the nature of core consciousness; how and why extended consciousness must be built on core consciousness; and why core consciousness turns on a feeling. Even if I can't answer the question, I hope to frame it clearly enough to see where we might look for an answer. I don't think it is in the heavens, but here on earth, among the birds and the bees.

The first thing we must do is dispose of the idea that consciousness is anything like it seems. It's not. For example, consciousness seems to be unitary, which is to say, not fractured into pieces. We don't have separate streams of consciousness running in our heads, but a single integrated perception, which nonetheless changes endlessly, shifting from second to second through a never-ending variety of states. Consciousness seems like a movie in the head, the pictures integrated not only with sounds but with smells, touches, tastes, emotions, feelings, thoughts, all bound to a sense of self, anchoring our whole being and its experiences to our body.

You don't have to think about this for long to realise that the brain must be binding sensory information together in some way, giving only a *perception* of a seamlessly integrated whole. Information from the eyes, the ears, the nostrils, from touch or memory or the bowels, enters different parts of the brain where it is processed independently, before ultimately giving rise to distinct perceptions of colour, aroma, touch, hunger. None of this is 'real'; it is all nerves firing; and yet we seldom mistake the objects we 'see' for aromas or sounds. While the retina does indeed form an inverted image of the world, the image is absolutely not seen on a movie screen by the brain, but is instead transformed into patterns of neurons firing down the optic nerve, more like a fax machine. Much the same happens when we hear or smell: nothing of the outside world gets into our head, only neurons firing. Likewise for bellyache; there's no reality but nerves.

For us to experience all of this consciously, moment by moment, as some kind of multimedia movie in the head, requires the conversion of all the digital dots and dashes into a perception of the 'real world', with all its sights and smells. And then, of course, we don't perceive this reconstructed world as being inside our heads, but we project it all back out again to where it belongs. We seem to view the world from a single cyclopean aperture in the front of the skull, which is quite obviously an illusion. Plainly all this involves a lot of neural jiggery-pokery. Equally plainly, wiring is important. Cut the optic nerves and you will be left blind. Conversely, stimulate the visual centres of the brain in a blind person, using an implanted array of microelectrodes, and they will see images generated directly by the brain, even if, so far, only rudimentary patterns. This is the basis of artificial sight, still very much in its infancy, but in principle feasible. By the same token, so is the film *The Matrix*, where all experiences are stimulated in a vat.

Just how much jiggery-pokery is going on is betrayed by the hundreds of strange, peculiar or outrageously weird cases from the annals of neurology, which exert a gossipy 'there but for the grace of God' fascination for most of us, an archive mined effectively, and tenderly, by Oliver Sacks and others. 'The man who mistook his wife for a hat' is probably Sacks's most celebrated case, even being adapted as a chamber opera by Michael Nyman, and later filmed. The man in question, identified only as 'Dr P.', was a musician of distinction, who suffered from a condition known as 'visual agnosia', in which his vision was perfectly intact but his ability to recognise and correctly identify objects, especially faces, became woefully inadequate. When in consultation with Sacks, he mistook his foot for his shoe, and later, when trying to take his hat, reached instead for his wife's head. The degeneration of a region of his brain responsible for visual processing (caused by an unusual form of Alzheimer's disease) reduced his visual world to meaningless patterns of abstract shapes, colours and movements, while sparing his cultivated mind and his wonderful musicianship.

This kind of degeneration is thankfully rare, but from the vantage point of neurology it is but one of a kind. A related condition, caused by damage to another restricted region of the brain, is known as Capgras syndrome. Here the patient can recognise people perfectly well, but has a bizarre belief that spouses or parents are not what they seem, but imposters who have taken their place. The trouble is not with people in general, but only with close friends and family, people who are emotionally close. In this case, the problem is in the nerves that connect the visual centres to the emotional centres of the brain (such as the amygdala). By severing these links, a stroke or other local injury (like a tumour) prevents the normal emotional response to seeing a loved one, a response that can be picked up using a lie detector. As the neurologist V. S. Ramachandran quips, even if you're not a good Jewish boy, seeing your mother makes your hands sweat. Sweat changes the electrical resistance of the skin, and it's this that is registered by a lie detector. But people with Capgras syndrome don't sweat when they see loved ones: their eyes tell them it's their mother, but their emotional centres fail to ground the impression. This emotional void appears to be the basis of the syndrome. The brain, conjuring with a lack of consistency, leaps to the absurd but logical notion that the person is an imposter. Emotions are more powerful than intellect, or better, are the foundation of intellect.

Cotard's syndrome is even stranger. Here the deficit is even more profound – all the senses become disconnected from the emotional centres of the brain, giving an emotional flatline. If everything perceived by all senses registers null on the emotion score, the bizarre, albeit again 'logical', conclusion, is that the person must be dead. Logic is twisted to accommodate emotion. A patient with Cotard's syndrome actually believes he is dead, and may even claim to smell rotting flesh. If asked, he will agree with you that the dead don't bleed;

but prick him with a needle and he'll look astonished, before conceding that the dead do bleed after all.[4]

My point is that specific brain injuries (lesions) cause specific *reproducible* deficits. It's hardly surprising, but a lesion in the same area causes the same deficit in different people, or for that matter in animals. In some cases, such lesions affect sensory processing, causing conditions such as motion blindness, another bizarre syndrome where the person can't perceive moving objects, but instead sees the world as if lit by disco strobe-lights, making it nearly impossible to judge the speed of moving cars or even just fill a wine glass. In other cases, similar lesions alter consciousness itself. Patients with transient global amnesia can't plan or remember, and are conscious only of the here and now. Patients with Anton's syndrome are blind but deny it. Patients with anosognosia report that they are fine but in fact have a serious deficit like paralysis: 'It's just having a rest, doctor.' Patients with pain asymbolia feel pain but don't experience its aversive quality – they don't hurt. And patients with blindsight are not conscious of being able to see (they are genuinely blind) but are nonetheless capable of pointing correctly to objects when asked. This last example, blindsight, has been demonstrated in macaque monkeys trained to respond when they see (or fail to see) an object. It's one of many parallels made transparent by the ingenuity of a rising generation of experimental psychologists studying consciousness in animals.

How peculiar are such deficits! Yet through the careful studies of neurologists over the last century or more, their reality, their reproducibility, their causes (cause in the limited sense of a lesion in a particular part of the brain) have been tracked down. Equally peculiar disconnections occur if specific bits of the brain are stimulated with electrodes. This has been done, mostly some decades ago, on hundreds of people suffering from severe untreatable epilepsy, which at its worst induces generalised seizures leading to a catastrophic loss of consciousness, and sometimes dementia or partial paralysis. Many patients undergoing neurosurgery for epilepsy have acted as willing, and fully conscious, guinea pigs, reporting their sensations verbally to the surgeon. And so we know that stimulating one particular area of the brain generates an overwhelming sense of depression, which dissipates as soon as the stimulation ceases; and that stimulating other parts can induce visions or bring to mind excerpts of music. Stimulating one spot reliably produces an out-of-body experience, giving a sense that the soul is floating somewhere near the ceiling.

More recently, a sophisticated box of tricks has been applied for similar ends, using a helmet that trains weak magnetic fields to induce electrical changes in specific brain regions without surgery. The helmet achieved notoriety in the mid-1990s, when Michael Persinger, at Laurentian University in Canada, began stimulating the temporal lobe (beneath the temples of the head), and found that he reliably (in 80 per cent of people) induced mystical visions, a sense of the presence of God or even the devil, in the room. It inevitably became known as the God helmet, although a team of Swedish researchers have queried the findings. With an enjoyable sense of mischief, the British TV science documentary programme *Horizon* packed off the celebrated atheist Richard Dawkins to Canada to try the God helmet in 2003. To their disappointment, no doubt, it didn't induce a transcendental experience in Dawkins. Persinger explained the failure by noting that Dawkins had scored badly on a psychological scale measuring proneness to temporal lobe sensitivity. That's to say, the religious parts of his brain are indifferent at the best of times. But the well-known writer and experimental psychologist Susan Blackmore was more impressed: 'When I went to Persinger's lab and underwent his procedures, I had the most extraordinary experiences I've ever had... I'll be surprised if it turns out to be a placebo effect,' she said. Persinger, incidentally, is at pains to point out that the physical induction of mystical experiences does not argue against the existence of God; there would still have to be some 'physical mechanism for the transmission of supernatural experiences'.

The point is that the brain, and equally the mind, is remarkably splintered into specialised regions. We are not at all conscious of these internal workings. This picture is borne out by the effect of mind-bending drugs, which again act on exquisitely precise targets. Hallucinogens like LSD, psilocybin (an ingredient of mushrooms) and mescaline (a compound found in some cacti), for example, all act on a specific type of receptor (serotonin receptors) found on specific neurons (pyramidal neurons) in particular regions of the brain (layer five of the cortex). As the neuroscientist Christof Koch, at Caltech in Pasadena, observes, they don't mess up the brain's circuits in some holistic manner. Likewise, many antidepressant or antipsychotic drugs have very specific targets. An implication is that consciousness, too, does not just emerge holistically as some sort of 'field' from the general workings of the brain, but is a property of the specific anatomy of the brain, albeit a number of regions collaborating in unison at any one time. It is only fair to say that there is little agreement about this kind of thing, even among neuroscientists, but I shall try to justify this view in the coming pages.

Vision is more complex than it seems, but we can gain no sense of this complexity through introspection, by 'thinking about' how or what we see; and it could never have been predicted by a philosopher thinking it out logically. Our conscious minds have no access to the neural mechanisms which underscore vision. The extent to which information is broken down into its constituent parts was barely conceived before the pioneering studies of David Hubel and Torsten Wiesel, at Harvard, from the 1950s on, work for which they were awarded the Nobel Prize (with Roger Sperry) in 1981. By inserting microelectrodes into the brains of anaesthetised cats, Hubel and Wiesel showed that different groups of neurons are activated by different aspects of the visual scene. We now know that every image is broken down into thirty or more channels, so some neurons fire only when they perceive an edge running in a certain direction, a diagonal or horizontal or vertical line, for example. Other cells fire in response to high or low contrast, or depth, or to a particular colour, or to motion in a particular direction, and so on. The spatial location of each such feature is also mapped out in relation to the visual field, such that a dark horizontal line in the top left corner of the visual field triggers several groups of neurons to fire, while a similar line in the bottom right corner stimulates a different group of neurons.

At each step the visual areas of the brain plot out a topographic map of the world. Only later, though, does this map accrue any meaning, the kind of meaning that eluded poor Dr P. To gain such meaning – Aha! It's a tiger! – visual information needs to be bound back together, almost certainly in several stages: lines and colours bound together into stripes, the broken outline of a crouching shape, and finally, drawing on experience, a full recognition of a tiger behind the bushes. Only the last stages of this process ever gain conscious representation; the vast majority of visual processing never sees the light of mind.

How are all these splintered bits of a scene bound back together again to produce a coherent vision? The question remains one of the most interesting in neurology, and has not yet been answered to everyone's satisfaction. In broad outline, though, the answer is that neurons fire in synchrony: neurons that fire together bind together. Precise timing is of the essence. Back in the late 1980s, Wolf Singer and colleagues, at the Max Planck Institute for Brain Research in Frankfurt, first reported a new type of brain wave, which could be picked up with an EEG (electroencephalogram), now known as gamma oscillations.[5] They found that large groups of neurons become synchronised to a common pattern, firing once every 25 milliseconds or so, which is to say, on average, around 40 times a second, or 40 hertz (Hz). (In fact there is a range between about 30 and 70 Hz, which is significant; we'll return to this later.)

Such synchronised patterns of firing were just what Francis Crick had been looking for. After his celebrated achievements in cracking the DNA code, Crick turned his formidable mind to the problem of consciousness. Working with Christof Koch, he was searching for some sort of firing pattern that could correlate with consciousness itself – what he dubbed the 'neural correlates of consciousness', or NCC for short.

Crick and Koch were acutely aware that much of what goes on in visual processing remains unconscious. That makes the question of consciousness even curiouser. All sensory input to the brain comes in the form of neurons firing, yet some types of neuron-firing are perceived consciously, so we become aware of a colour or a face, whereas other types are not (all that unconscious visual processing of lines, contrast, distance, and so on). What's the difference?

Crick and Koch reasoned that there's no way to tell the difference if we don't even know which types of neurons are associated with a conscious perception, and which ones are not. What they wanted to find was a specific group of neurons that begins firing the very moment the subject becomes conscious of something (of seeing a dog, for instance) and switch off as soon as our attention turns elsewhere. Presumably, Crick and Koch proposed, there should be something different about the firing of neurons that actually do give rise to conscious perceptions. Their quest for the neural correlates of consciousness became a kind of neurological Holy Grail. The 40 Hz oscillations caught their imagination because they offered, and still do, a conceptual answer. Neurons that fire together are linked across great swathes of the brain at any one moment. All the parallel circuitry is reduced to a serial output by time itself. And so consciousness varies from moment to moment as the instruments of an orchestra, the diverse melodic lines bound together in harmony at each moment. In the words of T. S. Eliot, you are the music while the music lasts.

The idea is beguiling but quickly gets more complicated when you think about it. The basic trouble is that binding has to occur in multiple levels, and not only within the visual system. Other aspects of the mind seem to work in much the same way. Take memory, for example. In his fine book *The Making of Memory*, the neurochemist Steven Rose recalls how baffled he'd been at the way in which memories dissipate like smoke all across the brain; they don't seem to be 'seated' in any one place. He later discovered that this is because they are broken down into their component parts in much the same way as vision. In newborn chicks pecking at flavoured beads, for example, Rose found that the chicks soon learnt to avoid acrid-tasting colour-coded beads, but that their memory is stored in pieces: colour in one place, shape in another, size, smell or acrid taste elsewhere, and so on. To form a coherent memory requires binding all these elements back together, in what amounts to a simulated rerun. And recent research duly shows that rebinding the constituents of memory relies on the firing of exactly the same ensemble of neurons that responded to the experience in the first place.

The neurologist Antonio Damasio goes further still, incorporating the 'self' into more neural maps. He discriminates carefully between emotions and feelings (some say too carefully). An emotion, for Damasio, is very much a physical bodily experience, bowels clenching with fear, heart pounding, palms sweating, eyes widening, pupils dilating, mouth grimacing, and so on. This is unconscious behaviour, largely beyond our control and even our ability to imagine, at least for those of us with a cushioned urban lifestyle. In all my years of climbing, there have been just two or three occasions when I was gripped with an animal fear, a bowel-churning emotional response that shocked me in its intensity. I smelt my own fear only once but I will not forget it: it is disturbing. For Damasio, all emotions, even the more graded emotions, are physical, set in the body. But the body is not separate from the mind: it is bound to the mind. All these body states feed back to the brain via nerves and hormones, and the change in body state is mapped out bit by bit, organ by organ, system by system. Much of this mapping takes place in the older parts of the brain, including the brain stem and midbrain, centres that are preserved faithfully in the brains of all vertebrates. These mind maps constitute *feelings*, the neural mappings of bodily emotions. How such neural mappings (basically information) give rise to a subjective sense of feeling is a moot point that we will return to in a little while.

But even feelings are not enough for Damasio: we are not conscious until we begin to feel our own feelings, to know the feeling. More maps, of course. So primary neural maps plot out our bodily systems – muscle tone, stomach acidity, blood sugars, breathing rate, eye movements, pulse, bladder distension, and so on – mapping and remapping every fleeting instant. Damasio sees our sense of self as arising from all these bodily reports, initially in the form of an unconscious protoself, essentially a consolidated readout of bodily status. True self-consciousness emerges from the way in which these mind maps are altered in relation to 'objects' in the outer world, objects like your son, that girl, a dizzying cliff, the smell of coffee, a ticket inspector, and so on. All these objects are perceived directly by the senses, but also generate an emotional response in the body, which is picked up by the neural body maps in the brain to generate a feeling. Consciousness, then, is the knowledge of how objects in the world alter the self: a map of all these maps and how they change, which is to say, a second-order map. A map of how feelings relate to the world. A map that drapes our perceptions with values.

How are all these maps built? And how do they come to relate to each other? The most persuasive answer comes from the neuroscientist Gerald Edelman, who, after winning the Nobel Prize for his contributions to immunology in

1972, dedicated the following decades to work on consciousness. His ideas draw from the same well as his work in immunology, the power of selection within the body. In the case of immunology, Edelman helped show how a single antibody can be selectively amplified after contact with a bacterium: selection leads to a proliferation of the winning immune cells at the expense of others. After half a lifetime, the specificity of immune cells in your blood-stream depends largely on your experiences, not on your genes directly. According to Edelman, a similar type of selection operates continuously in the brain. Here, groups of neurons are selected and strengthened through use, while other groups wither away through disuse. Again, the winning combinations dominate. And again, the relationships between neurons are determined by accumulated experience, not specified by genes directly.

It happens like this. During embryonic development, the brain is wired up in a rough-hewn 'chunky' manner, with bundles of nerve fibres connecting different regions of the brain (the optic nerve to the visual centres, the corpus callosum connecting the two hemispheres, and so on) but with little in the way of specificity or meaning. In essence, genes specify the general circuitry of the brain, whereas experience specifies the exact wiring and all the idiosyncratic detail that entails. Meaning comes mostly with experience, which is written directly into the brain. As Edelman puts it, 'neurons that fire together wire together'. In other words, neurons that fire at the same time strengthen their connections (synapses) and form more connections that bind them together physically.[6] Such connections form locally within groups of neurons (helping bind together different aspects of visual information, for example), but also over great distances, connecting the visual centres with emotional or speech centres. At the same time, other synaptic connections grow weaker, or disappear altogether, as their neurons have little in common. Soon after birth, as the flow of experience gathers pace, the mind is sculpted from within. Billions of neurons die: somewhere between 20 per cent and 50 per cent of all neurons are lost within the first months of life, and tens of billions of weak synaptic connections are lost. At the same time tens of *trillions* of synapses are strength-ened, generating as many as 10,000 synapses per neuron in some regions of the cortex. Such synaptic plasticity is greatest in our formative years, but con-tinues throughout life. Montaigne once said that every man over the age of forty is responsible for his own face. We are undoubtedly responsible for our own brain.

You may be wondering how exactly the genes contribute to this process. They do so not only by specifying the general architecture but also the relative size and development of different brain regions. They affect the likelihood of survival of neurons, the strength of their synaptic connections, the proportion of excitatory to inhibitory neurons, the overall balance of distinct neurotransmitters, and so on. Such influences plainly contribute to our personality, as well as our liability to become addicted to dangerous sports, or drugs, or to fall into deep depression, or think rationally, and in this way genes influence gifts and experience too. But genes do not specify the detailed neural architecture of the brain. How could they? There is no way that 30,000 genes could determine the 240 trillion (a trillion is a thousand billion) synapses of the cerebral cortex (Koch's figure). That works out at one gene per 8 billion synapses.

Edelman refers to the process of brain development as *neural darwinism*, which emphasises the idea that experience selects successful neural combinations. All the basic tenets of natural selection are present: we start out with a massive population of neurons, which can be wired up in millions of different ways to achieve the same ends. The neurons vary among themselves and can either grow more robust or wither away; there is competition between neurons to form synaptic connections and differential survival on the basis of success, the 'fittest' neuron combinations forming the most synaptic connections. Francis Crick famously jibed that the whole construction is better termed 'neural edelmanism', for the parallels with natural selection are a bit forced. Nonetheless, the basic idea is now widely accepted by neurologists.

Edelman made a second important contribution to the neural basis of consciousness: the idea of reverberating neural loops, or what Edelman calls (a little unhelpfully) parallel re-entrant signals. He means that neurons firing in one region connect with neurons in faraway regions, which reciprocate through other connections to form a transient neural circuit, reverberating in synchrony before competing sensory inputs dissolve the ensemble of neurons and replace it with another fleeting ensemble, again reverberating in unison. Here, the ideas of Edelman mesh beautifully with those of Crick and Koch, and Wolf Singer (though it must be said that one often has to read between the lines to appreciate their commonalities; I have rarely come across a field where the main protagonists refer so little to each other, rarely even stooping to condemn the misguided ideas of their opponents).

Consciousness operates on a scale of tens to hundreds of milliseconds.[7] Flash up two images momentarily, with a 40 millisecond gap between them, and you will be consciously aware of only the second image; you simply won't see the first image. And yet microelectrodes and brain scans (such as functional magnetic resonance) show that the visual centres of the brain did pick up the first image; it just never became conscious. To gain consciousness, it seems likely that a neural cohort has to reverberate together for tens, perhaps hundreds, of milliseconds; back again to Singer's 40 Hz oscillations. Both Singer and Edelman have shown that distant regions of the brain do indeed oscillate in synchrony in this way: they 'phase lock' together. Other groups of neurons lock together in different phases, slightly faster or slower. In principle, such phase-locking enables discrimination between different aspects of the same scene. So all the elements of a green car phase-lock together, while those of a blue car nearby phase-lock slightly differently, ensuring that the two cars don't get muddled up in the mind. Each aspect of a visual scene phase-locks slightly differently.

Singer has a lovely idea that explains how all these phase-locked oscillations bind together at a higher level, at the level of consciousness itself: that's to say, how these oscillations bind to other sensory inputs (hearing, smell, taste, and so on) and to feeling, memory and language, to generate an integrated sense of consciousness. He calls it *neural handshaking*, and it allows the hierarchical 'nesting' of information, so that smaller bits of information find their place in the bigger picture. Only the top hierarchy, which amounts to a kind of executive summary of all the non-conscious information, is consciously perceived.

Neural handshaking depends on a simple fact: when a neuron fires, it depolarises, and it can't fire again until it has repolarised. That takes some time. And this means that if another signal arrives during the fallow repolarizing period, it will be ignored. Consider: if a neuron is firing 60 times a second (60 Hz) it is constrained to receive signals only from other neurons that are firing in phase synchrony. So, for example, if a second group of neurons happens to be firing 70 times a second (70 Hz) they will be out of synchrony with the first group most of the time. They become independent units, unable to shake hands. On the other hand, if a third group of neurons is firing more slowly, say at 40 Hz, there is a much longer period when these neurons are repolarised and ready to fire, merely waiting for the right trigger. Such neurons can fire in response to neurons oscillating at 70 Hz. In other words, the slower the oscillation, the greater the phase overlap, and the better the possibility of handshaking with other groups of neurons. So, the fastest oscillations bind together discrete aspects of the visual scene, smells, memory, emotions, and so on, each as independent units, whereas slower oscillations bind together all the sensory and bodily information into a unified whole (Damasio's second-order map) – a moment in the flow of consciousness.

Little of this is proved beyond doubt, but there is plenty of evidence that is at least consistent with this picture. Most importantly, these ideas make testable predictions, for example, that 40 Hz oscillations are necessary to bind the contents of consciousness and, conversely, that a loss of such oscillations equates to a loss of consciousness. Given the difficulty of making such measurements (which requires measuring the firing rates of thousands of individual neurons across the brain, simultaneously) it may be years before these hypotheses, or others, are verified.

Even so, as an explanatory framework, these concepts help make consciousness intelligible. For example, they show how extended consciousness can develop from core consciousness. Core consciousness operates in the present, rebuilding itself moment by moment, mapping out how the self is altered by external objects, draping perceptions with feelings. Extended consciousness uses the same mechanisms, but now binds memories and language into each moment of core consciousness, qualifying emotional meaning with an autobiographical past, labelling feelings and objects with words, and so on. Thus extended consciousness builds on emotional meaning, integrating memory, language, past and future, into the here and now of core consciousness. The selfsame neural handshaking mechanisms allow a vast expansion of parallel circuitry to be bound back into a single moment of perception.

I find all this believable. And yet still the deepest question of all remains

unanswered. How do neurons generate feeling in the first place? If consciousness is the ability to feel a feeling, to generate nuanced emotional meaning, a
running commentary on the self in the world, the entire edifice dances on that
pinhead of feeling: what philosphers call the problem of qualia. It's time to
face the hard problem head on.

Pain hurts for a reason. A few unfortunate people are born with a congenital
insensitivity to pain. They suffer from terrible, often unanticipated afflictions.
One four-year-old girl, Gabby Gingras, was the subject of a documentary
film directed by Melody Gilbert in 2005. Without pain, each developmental
milestone became an ordeal. When her milk-teeth first cut through, Gabby
chewed her own fingers to the bone, mutilating them so badly her parents
were forced to have her teeth removed. On learning to walk, Gabby injured
herself time and again, on one occasion breaking her jaw without knowing it,
until an infection caused a fever. Worse, she would poke herself in the eye,
causing damage that required stitches, which she soon tore out. Her parents
tried restraints, then goggles, to no avail. At the age of four, she had to have
her left eye removed; and her right eye too is seriously damaged, leaving her
legally blind (20/200 eyesight). As I write, Gabby is now seven, and coming
to terms with her predicament. Others like her have died in childhood; a few
survive to be adults, but often have to contend with serious injuries. Gabby's
parents started a foundation, Gift of Pain, which supports people with the
condition (thirty-nine members so far). It's well named: pain is undoubtedly
a blessing.

Pain is not alone. Hunger, thirst, fear, lust, these are among what Derek
Denton calls the 'primordial emotions', which he describes as imperious sensations that commandeer the whole stream of consciousness, compelling
intentions to act. All are obviously tailored to the survival of the organism, or
to its propagation: the feeling forces the act, and the act in turn saves, or
propagates, life. Humans may well be alone in having sex knowingly to breed;
but even the Church has had little success in eliminating gratification from
copulation. Animals, and most humans, copulate for the reward of orgasm,

not for the sake of offspring. The point is that all these primordial emotions are *feelings*, and all serve a biological purpose, even if that purpose is not always appreciated. Pain is, above all else, an unpleasant feeling. Without the agonising *pain* of pain, we would injure ourselves horribly; experiencing pain without its aversive quality is no use. Likewise lust. Mechanical copulation is no reward in itself: we, all animals, seek the rewards of the flesh, the *feeling* of it. And again, simply registering thirst on a neural dial in the desert is not sufficient: it is the raging emotion that consumes the mind from within that aids our survival, that forces us onwards to an oasis, that wrings out one final drop of stamina.

The idea that such primal emotions evolved by natural selection is hardly challenging, but it has significant connotations, first pointed out by the founding father of modern psychology, the American genius William James, in late Victorian times. James argued that feelings, and therefore by extension consciousness itself, have biological utility. This, in turn, means that consciousness is no 'epiphenomenon', accompanying the organism around like a shadow but unable to exert any physical effects on its own. Feelings *do* exert physical effects. And yet if feelings do exert physical effects, then they must in some sense *be* physical. James concluded that despite their unphysical appearance, feelings are indeed physical and evolved by natural selection. But what actually *are* they? Nobody ever pondered that harder than James himself, and the conclusion he came to was counterintuitive and troubling. There must exist unknown properties of matter, he argued, some sort of 'mind-dust' that permeates the universe. Despite being the averred hero of many distinguished neurologists today, James embraced a form of pan-psychism (consciousness is everywhere, part of everything) that few have been inclined to follow. Until now.

Just to give a sense of how difficult this problem is, consider a few gadgets like the television, fax machine or telephone. You don't need to know how any of these work to appreciate that they don't break the laws of physics. Electronic signals encode the output in one way or another, but the output is always physical: patterns of light for a TV, sound waves in the case of a telephone or radio, print for fax machines. An electronic code elicits an output in a known physical medium. But what about feelings? In this case, the nerves transmit coded electronic signals in essentially the same way as a TV: neurons specify the output precisely, via some code. No problem there. But what exactly is this output? Think about all the known properties of matter. Feelings don't seem to be electromagnetic radiation or sound waves, or to correspond to anything in the known physical structure of atoms. They're not quarks, they're not electrons; what on earth are they? Vibrating strings? Quantum gravitons? Dark matter?[8]

This is the 'hard problem' enunciated by Chalmers; and, like James before him, Chalmers too has argued that it can only be answered through the discovery of new, fundamental, properties of matter. The reason is simple. Feelings are physical, yet the known laws of physics, which can supposedly give us a complete account of the world, have no place for them. For all its marvellous power, natural selection doesn't conjure up something from nothing: there has to be a germ of *something* for it to act upon, a germ of a feeling, you might say, that evolution can fashion into the majesty of mind. This is what Scottish physical chemist Graham Cairns-Smith calls 'the bomb in the basement' of modern physics. Presumably, he says, if feelings don't correspond to any of the known properties of matter, then matter itself must have some additional features, 'subjective features', that when organised by selection ultimately give rise to our inner feelings. Matter is conscious in some way, with 'inner' properties, as well as the familiar external properties that physicists measure. Pan-psychism is taken seriously again.

It sounds preposterous. But what arrogance to think that we know all there is to know about the nature of matter! We don't. We don't even understand how quantum mechanics works. String theory is sublime in the way that it spins out the properties of matter from the vibrations of unimaginably small strings in eleven unimaginable dimensions; but we have no way of determining experimentally whether there's any grain of truth in it. And that's why I noted at the beginning of this chapter that the Pope's position was not unreasonable. We don't know enough about the deep nature of matter to know how neurons transform brute matter into subjective feelings. If electrons can be both waves and particles at the same time, why shouldn't spirit and substance be two aspects of the same thing?

Cairns-Smith is better known for his work on the origin of life, but since retiring has turned his fine mind to the problem of consciousness. His books on the subject are penetrating and entertaining, and follow the likes of Roger Penrose and Stuart Hameroff into the quantum gardens of mind. Cairns-Smith sees feelings as the coherent vibrations of proteins. They are coherent in the same sense that a laser beam is coherent, which is to say that the vibrations (phonons) coalesce into the same quantum state. This is now a 'macro-quantum' state, writ large across great tracts of the brain. Once again, Cairns-Smith evokes an orchestra, where the vibrations of individual instruments coalesce into transcendent harmonies. Feelings are music, and we are the music while the music lasts. It's a beautiful conception. Nor is it unreasonable to ascribe quantum effects to evolution. There are at least two cases where the blind forces of selection may well have recruited quantum mechanisms: the passage of light energy to chlorophyll in photosynthesis, and the tunnelling of electrons to oxygen in cellular respiration.

And yet I don't quite buy it for the mind. The quantum mind may exist, but there are several problems that together look insurmountable to me.

The first and most important is logistical. How do quantum vibrations hop across synapses, for example? As Penrose himself acknowledges, a macro-quantum state that is restricted to the inside of single neurons solves nothing; and on the quantum level a synapse is an ocean. For phonons to vibrate in concert there must be a repeating array of proteins, spaced closely together so each vibrates in unison before the phonons decay. Such questions can be tackled experimentally, but as yet there is not a jot of evidence that coherent macroquantum states exist in the mind. Quite the contrary: the brain is a hot, wet and turbulent system, about the worst possible place to generate a macro-quantum state.

And if the purported quantum vibrations really do exist, and do depend on repeating arrays of proteins, then what happens to consciousness if these protein arrays are disrupted by neurodegenerative disease? Penrose and Hameroff impute consciousness to the microtubules inside neurons but these degenerate in Alzheimer's disease, ending up as the tangles that are a classic sign of the disease. Such tangles are found in their thousands early in the course of the condition (mostly in the parts of the brain responsible for forming new memories), and yet conscious awareness remains intact until quite late in the disease. There's no correlation, in short. Much the same is true of other postulated quantum structures. Myelin sheaths for example, that envelop neurons in the white matter, are stripped away in multiple sclerosis, again with little or no impact on consciousness. The only example that is at least consistent with a quantum explanation is the behaviour of support cells called astrocytes after a stroke. In one study, several patients were not consciously aware of their own recovery after a stroke: there was an odd gap between their measured performance and their own perception of that performance, which might (or might not) be explained in terms of quantum coherence across the astrocyte network (if, indeed, such a network even exists, which now looks doubtful).

A second question relating to quantum consciousness is what the concept actually solves, if anything. Let's assume that there really is a network of vibrating proteins in the brain, which 'sing' in unison, their melodies giving rise to feelings, or rather, *being* feelings. Let's assume too that these quantum vibrations somehow 'tunnel' through the oceanic synapses and trigger another quantum 'song' on the other side, spreading the coherence across the brain. What we have here is a whole parallel universe in the brain, one which must operate hand in hand with the known 'classical' universe of neurons firing, for how else could synchronised firing give rise to conscious perceptions, or neurotransmitters influence conscious state, which they most certainly do? The quantum universe would need to be compartmentalised in exactly the same way as the brain. So feelings related to vision (like seeing red) would need to be restricted to the visual processing areas, whereas emotional feelings could vibrate only in regions like the amygdala or the midbrain. One problem with this is that the microscopic infrastructure in all neurons is basically the same – the microtubules in one neuron don't differ in any meaningful way from those in any other neuron – so why should some sing in colour, and others in pain? Hardest of all to swallow is the fact that feelings echo the basest of all bodily concerns. One could perhaps imagine a fundamental property of matter resonating love or music, but stomach ache? Is there a unique vibration that simply *is* the feeling of bladder distension in public? Improbable. If God plays dice, surely that's not the game. But if not quanta, then what?

Where might we best look for an answer to the 'hard problem' of consciousness? Several of the apparent paradoxes can be dealt with quite simply, including Cairns-Smith's 'bomb in the basement' of modern physics. Do feelings really have to be a physical property of matter if they are to evolve by natural selection? Not necessarily. Not if neurons encode feelings in an exact and reproducible manner, which is to say, if a group of neurons firing in a certain way *always* gives rise to the same feeling. Then selection simply acts on the underlying physical attributes of neurons. Edelman, careful as ever in his choice of words, prefers the word 'entails'. A pattern of neurons firing *entails* a feeling; it is inseparable from it. In the same way, you might say that a gene entails a protein. Natural selection acts on the properties of proteins, not the sequences of genes, but because genes encode proteins in a strict fashion, and because only the gene is inherited, it amounts to the same thing. Certainly, it seems to me far more likely that primordial emotions, such as hunger and thirst, should be entailed by an exact pattern of neural firing, rather than being some fundamental vibrational property of matter.

Another paradox that can be addressed quite simply, at least in part, is the perception that our minds are immaterial, and our feelings ineffable. According to another fine scientist who has turned his mind to consciousness in retirement, the New York physician and pharmacologist José Musacchio, the essential insight is that the mind does not, indeed cannot, detect the existence of the brain. We perceive neither the brain nor the physical nature of the mind by thinking about it. Only the objective methods of science have linked the mind with the physical workings of the brain. How remarkably misguided we have been in the past is exemplified by the ancient Egyptians, who in embalming their kings preserved the heart and other organs with great care (they took the heart to be the seat of emotion and mind), but scooped the brain out through the nose with a hook, cleaning the cavity with a long spoon, and flushed away the mess. They were uncertain what the brain was for, and assumed it would not be needed in the next world. Even today, we get a sense of the peculiar inability of the mind to detect itself during brain surgery. Sensitive to so much of the world, the brain has no pain receptors of its own, and so is totally insensitive to pain. This is why neurosurgery can be performed without general anaesthetic.

Why should the mind not detect its own physical workings? It is plainly disadvantageous for an organism to cogitate on its own mind when it should really be focusing all its brainpower on detecting a tiger in the bushes, and deciding what to do about it. Introspection at inopportune moments is not a property likely to survive the rigours of selection. The outcome, though, is that our perceptions and feelings are transparent: they are simply there, without any whiff of their physical neural basis. Because we are necessarily unaware of the physical basis of our perceptions and feelings, the conscious mind has a strong sense of immateriality, or spirituality. It might be a troubling conclusion for some people, but it seems inescapable: our sense of spirituality arises from the fact that consciousness operates on a need-to-know-only basis. We're shut out of our own brain for the sake of survival.

Much the same goes for the ineffability of feelings. If, as I've argued, feelings are entailed by patterns of neural firing, by a very precise code, then feelings amount to quite a sophisticated non-verbal language. Verbal languages are deeply rooted in this non-verbal language, but can never *be* the same thing. If a feeling is entailed by a neural pattern, the word for that feeling is entailed by a different neural pattern: it is translated from one code into another, one language into another. Words can only describe feelings through translation, and so feelings are strictly ineffable. And yet all of our languages are anchored in shared feelings. The colour red doesn't really exist. It is a neural construct that can't be conveyed to someone who has not perceived something similar themselves. Likewise the feeling of pain or hunger, or the smell of coffee, all of them are sensations which anchor words and make verbal communication possible. As Musacchio notes, there always comes a point when we are obliged to say, 'Do you know what I mean?' Because we share the same neural structures and feelings, languages are grounded in our common human experience. Language without feeling is bereft of meaning, but feelings exist, meaning exists, without any verbal language, as a core consciousness of mute emotions and wordless perceptions.

All this means that, however feelings may be generated by neurons, we'll never get close to it through introspection or logic – through philosophy or

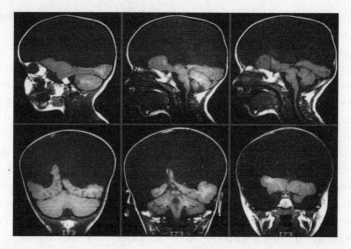

Figure 9.1 MRI scan of the head of a child with hydranencaphaly. Remarkably, virtually all the cerebral cortices are missing, and the cranium is instead filled with cerebrospinal fluid.

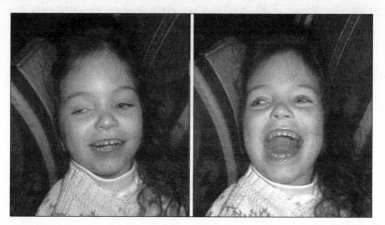

Figure 9.2 Happiness and delight on the face of a four-year-old girl, Nikki, with hydranencephaly.

theology – but only by experiment. On the other hand, the fact that consciousness is grounded in feelings, motivations and aversions means that we can get at the roots of consciousness without needing to communicate verbally with other animals: we just need clever experimental tests. And that in turn means we should be able to study the critical neural transform, from firing to feeling, in animals, even in simple animals, for all the signs are that the primordial emotions are widespread in vertebrates.

One remarkable suggestion that consciousness is more widespread than we like to credit is the survival and apparent consciousness of those few exceptional children who are born without cerebral cortices (see Fig. 9.1). A small stroke, or similar developmental abnormality, can lead to the reabsorption of large parts of both cortices during pregnancy. It's hardly surprising that such children are born with many handicaps, lacking language and good sight, but according to the Swedish neuroscientist Björn Merker, despite the absence of nearly all the brain regions that we normally associate with consciousness, some of these children are capable of emotional behaviour, laughing and crying appropriately, and showing signs of genuinely human expression (see Fig. 9.2). I mentioned earlier that many emotional centres of the brain are located in ancient parts of the brain, the brain stem and midbrain, which are shared by almost all vertebrates. Through MRI scans, Derek Denton has shown that these ancient areas mediate the experience of primordial emotions such as thirst and the fear of suffocation. It may well be that the roots of consciousness are not to be found in the new-fangled cortices at all, which of course *elaborate* consciousness enormously, but in the ancient and densely organised parts of the brain shared widely with many other animals. And if that's the case, then the neural transform, from firing to feeling, loses some of its mystique.

Just how widespread is consciousness? Until we come up with some sort of consciousometer we'll never know for sure. And yet the primordial emotions – thirst, hunger, pain, lust, horror of suffocation, and so on – all of these seem to be widespread among animals that have brains, including even simple invertebrates like bees. Bees have less than a million neurons (we have 23 *billion* in the cortex alone), and yet are capable of quite sophisticated behaviour, not only signalling the direction of food through their famous waggle dance, but even optimising their behaviour to visit primarily the flowers that are the most reliable source of nectar, even when the nectar balance is wilfully modified by wily researchers. I would not argue that bees are conscious in the way that we understand the term, yet even their simple neural 'reward system' demands a reward, which is to say, a *good feeling*, that sweet taste of nectar. Bees, in other words, already have what it takes to be conscious, even if they might not yet be truly conscious.

And so feelings are ultimately a neural construct, and not a fundamental property of matter. In some parallel universe, where the highest evolutionary attainment was the bee, would we really feel compelled to reach for new laws of physics to explain their behaviour? But if feelings are no more than neurons doing their thing, why do they seem so real, why *are* they so real? Feelings *feel* real because they have real meaning, meaning that has been acquired in the crucible of selection, meaning that comes from real life, real death. Feelings are in reality a neural code, yet a code that is vibrant, rich in meaning acquired over millions or billions of generations. We still don't know how our neurons do it, but consciousness, at bottom, is about life and death, and not the wonderful pinnacles of the human mind. If we really want to understand how consciousness came to be, then we must remove ourselves from the frame.

DEATH

The Price of Immortality

It's said that money can't buy happiness. But Croesus, King of Lydia in ancient times, was rich as ... Croesus, and thought himself the happiest of men. Seeking avowal from the Athenian statesman Solon, then passing through his lands, Croesus was irritated to be told, 'Count no man happy until he be dead'; for who can predict what Fate holds in store? And it so happened that Croesus, acting on an archetypically ambiguous oracle from Delphi, was captured by Cyrus, the Great King of Persia, and bound to a pyre to be burned alive. Yet instead of berating the gods for his excruciating end, Croesus murmured the name 'Solon'. Mystified, Cyrus enquired what he had meant, and was told of Solon's counsel. Realising that he, too, was a puppet of fortune, Cyrus had Croesus cut down (others say that Apollo came to his aid with a thunderstorm) and appointed him as an adviser.

Dying well meant a great deal to the Greeks. Fate and death were played out by invisible hands, which intervened in the most involved ways to bring men to their knees. Greek theatre is full of tortuous devices, death preordained by the fates, prefigured in cryptic oracles. As in frenzied Bacchic rituals and the fables of metamorphosis, the Greeks seem to owe something of their fatalism to the natural world. And vice versa: from the perspective of Western culture, the elaborate deaths of animals sometimes seem to assume the shade of Greek theatre.

There's more than an element of Greek tragedy, for example, about the mayflies, which live for months as larvae, before metamorphosing into adults lacking in mouthparts and digestive tract. Even the few species that live out their single orgiastic day are fated soon to starve. What about the Pacific salmon, which migrate hundreds of miles to the streams of their birth, wherein their hormonally charged frenzy is cut short by a catastrophic demise and death within days? Or the Queen bee, who shows no sign of her age for sixteen years, until finally her supply of sperm runs dry, whereupon she is torn apart by her own daughters? Or the twelve-hour copulation frenzy of the Australian marsupial mouse, culminating in death by depression and exhaustion, which can be prevented by castration? Tragedy or comedy, this is certainly dramatic. These animals are as much the pawns of fate as Oedipus himself. Death is not only inevitable; it is controlled by the fates, programmed into the very fabric of life.

Of all these grotesque modes of death, perhaps the most tragic, and resonant to us today, befell the Trojan Tithonus, whose goddess lover asked Zeus to confer immortality on him, but forgot to mention eternal youth. Homer has it that 'loathsome old age pressed full upon him', leaving him babbling endlessly. And Tennyson pictures him looking down on the 'dim fields about the homes of happy men that have the power to die, and grassy barrows of the happier dead'.

There is a tension between these forms of death, between the urgent death programmed into the lives of some animals and the abandonment of old age that faces humanity alone, the lack of a programme, the unspeakable endless end of Tithonus. This is exactly what we are visiting upon ourselves today, as medicine marches on, prolonging our lifespan but not our health. For every year of life granted by the gods of modern medicine, but a few months are spent in good health, and the rest in terminal decline. Like Tithonus, finally we beg for the grave. Death may seem a cruel cosmic joke, but ageing is mirthless.

Yet there should be no need to unmask Tithonus in our twilight years. Certainly, the intractable laws of physics forbid eternal youth as firmly as perpetual motion, but evolution is surprisingly flexible and shows that longer life is usually coupled with longer youth, avoiding the misery of Tithonus. Examples abound of animals whose lives have been extended painlessly, which is to say, without diseases, to two, three, even four times their original length, when circumstances change. One spectacular example is the brook trout that were introduced into the cold nutrient-poor waters of a lake in the Sierra Nevada in California. Their lifespan quadrupled from barely six years, to more than twenty-four, the only apparent 'cost' being a delay in sexual maturation. Similar findings have been reported among mammals such as opossums. When shielded from predation on islands for a few thousand years, for example, opossums more than double their normal lifespan, and age at half the rate. We humans, too, have doubled our maximal lifespan over the last few million years without any obvious penalty. From an evolutionary point of view, Tithonus ought to be a myth.

But mankind has sought eternal life for millennia and signally failed to find it. While advances in hygiene and medicine have prolonged our *average* lifespan, our *maximum* lifespan, at about 120 years, has remained stubbornly fixed, despite all our efforts. At the very dawn of recorded history, Gilgamesh, King of Uruk, sought everlasting life in the form of a fabled plant, which after an epic search slipped through his fingers like a myth. It's been the same ever since. The elixir of life, the holy grail, the ground horn of a unicorn, the philosophers' stone, yoghurt, melatonin, all have been purported to extend life; none has done so. Blatant charlatans rub shoulders with scholars to colour the history of rejuvenation research. The celebrated French biologist Charles Brown-Séquard injected himself with extracts from the testicles of dogs and guinea pigs, and reported improved vigour and mental powers to the Société de Biologie in Paris in 1889, even demonstrating the proud arc of his urine before the flabbergasted assembly. By the end of that year, some 12,000 physicians were administering Brown-Séquard's fluid. Surgeons around the world were soon implanting sliced testicles from goats, monkeys, even prisoners. Probably the most notorious American charlatan of all, John R. Brinkley, made a vast fortune from his goat-gland transplants, before dying a broken man, the victim of a thousand ungrateful lawsuits. It's doubtful whether mankind has added a single day to our allotted tenure on this earth, for all our overweening ingenuity.

So there is an odd gap between the flexibility of evolution – the ease with which lifespan seems to be moulded – and the blank intransigence that greets our efforts to today. How does evolution extend lifespan so easily? It's plain from our millennia of abject failure that until we understand the deeper reasons for death, we will never get anywhere. On the face of it, death is a perplexing 'invention': natural selection normally acts at the level of individual organisms, and it's hard to see how my death will benefit me, or what Pacific salmon gain from falling to pieces, or black widow spiders from being cannibalised. But it is equally plain that death is far from accidental, and it certainly evolved for the benefit of individuals (or rather, their selfish genes, in Richard Dawkins's unforgettable phrase) soon after the dawn of life itself. If we want to better our end, to evade the woes of Tithonus, we'd better go back to the beginning.

Imagine landing a time machine 3 billion years ago in shallow coastal waters. The first thing you notice is that the sky is not blue, but a dull hazy red, a touch reminiscent of Mars. The quiet seas reflect in shades of red. It's pleasantly warm in this haze, if too misty to see the sun clearly. On land not much catches the eye. The rocks are bare, with damp patches of discoloration here and there, bacteria clinging precariously to their extreme terrestrial outpost. There's no grass or vegetation of any sort. But marching through the shallow waters are scores of strange domes of greenish rock. Apparently the work of life, the tallest are a metre or so high. A few similar rare structures are still found in the most remote and inaccessible bays on earth today: stromatolites. Nothing else stirs in the waters. There's no fish here, no seaweed, no scuttling crabs, no waving anemones. Take off your oxygen mask, and you'll soon understand why: you'll suffocate in a few minutes. There's barely any oxygen, even close to the stromatolites. Yet their blue-green bacteria, *cyanobacteria*, are already beginning to lace the air with traces of that hazardous gas. In a billion years, their emissions will finally turn our planet vivid greens and blues. And only then will we recognise this bare place as home.

Seen from space, if we could penetrate the dull red smog, there's only one feature of this early earth that is more or less the same today: algal blooms. These too are cyanobacteria, related to those in stromatolites, but floating in vast plaques at sea. From space they look much the same as modern algal blooms; and under the microscope, the ancient fossils are virtually identical to modern cyanobacteria, like *Trichodesmium*. The blooms persist for weeks, their rapid growth stimulated by minerals carried out to sea from rivers, or dredged up from the ocean depths by rising currents. Then overnight they vanish, dissolving in the water, which once more reflects only the lifeless red sky. Today, too, vast ocean blooms dissolve overnight without warning.

It's only recently that we've come to appreciate what's going on. These vast hordes of bacteria don't just die: they kill themselves quite deliberately. Each and every cyanobacterium contains within itself the machinery of death, an ancient system of enzymes remarkably similar to those in our own cells, dedicated to dismantling the cell from within. It's so counterintuitive, the idea of bacteria liquidating themselves, that researchers have tended to overlook the evidence; but it is now too strong to ignore. The fact is that bacteria die 'deliberately', and the genetic evidence, marshalled by Paul Falkowski and Kay Bidle at Rutgers University, New Jersey, implies that they've been doing so for three billion years. Why?

Because death pays. Bacterial blooms are made up of countless trillions of genetically similar, if not identical cells. But genetically identical cells are not always the same. Just think of our own bodies, with several hundred different cell types, all of them genetically identical. Cells develop differently, or *differentiate*, in response to subtly different chemical cues from the environment, which in our case means the surrounding cells. In the case of bacterial blooms, the environment includes other cells, some of which emit chemical cues or even straight toxins, as well as physical stresses, such as the level of sunlight, nutrient availability, viral infection, and so on. So they might be genetically identical but their environment batters them in endlessly inventive ways. And that is the basis of differentiation.

Three billion years ago we see the first signs of differentiation – genetically identical cells are beginning to take on diverse appearances, appointed to different fates, depending on their life history. Some become hard resistant spores, others form into thin sticky films (biofilms) that attach on to submerged surfaces like rocks; some flourish independently, away from the tribe, and others simply die.

Or rather, they don't simply die: they die complicatedly. Quite how the complex machinery of death first evolved is uncertain. The most plausible answer is through interactions with phages, a type of virus that infects bacteria. Viral particles are found in modern oceans in shocking abundance: hundreds of millions of them per millilitre of seawater, which is at least two orders of magnitude more than bacteria; and they almost certainly existed in comparable numbers in ancient times. The unceasing war between bacteria and phage is one of the most significant and unsung forces in evolution. Programmed death likely emerged as one of the earliest weapons in this war.

A simple example is the toxin–antitoxin modules employed by many phages. Among their few genes, these phages encode some toxin, capable of killing the host bacteria, along with an antitoxin that protects the bacteria against the toxin. In dastardly manner, the toxin is long-lasting, while the antidote is short-lived. Infected bacteria produce both the toxin and antitoxin, and so survive, while naive bacteria or infected bacteria that attempt to cast off the viral shackles are vulnerable to the toxin and die. The simplest escape route for the hapless bacteria is to grab the gene for the antitoxin and incorporate that into its own genome, so affording it protection even if not infected. And so the war propagates, with more complex toxins and antitoxins evolving, the machinery of war growing ever more baroque. This may well be the way in which *caspase* enzymes first evolved, perhaps in cyanobacteria.[1] These specialised 'death' proteins cut up cells from within. They act in cascades, in which one death enzyme activates the next in the cascade, until a whole army of executioners is let loose upon the cell.[2] Importantly, each caspase has its own inhibitor, an 'antidote' capable of blocking its action. The whole system of toxins and antitoxins, cobbled together into multiple levels of attack and resistance, might well betray the long drawn-out evolutionary battle between phage and bacteria.

While such battles between bacteria and viruses probably lie at the deep roots of death, suicide undoubtedly benefits bacteria even in the absence of infection. The same principles apply. Any physical threat with the potential to wipe out the whole bloom (like intense ultraviolet radiation or nutrient deprivation) can trigger programmed cell death in cyanobacterial blooms. The most robust cells survive the threat by developing into hardy spores that seed the next bloom, while their more fragile, if genetically identical, siblings respond to the same threat by triggering the machinery of death. Whether these dynamics are best seen as murder or suicide is a matter of taste. Dispassionately, the outcome is simply that more copies of the bacterial genome survive over evolutionary time if damaged cells are eliminated. It's the simplest form of differentiation, a straight binary choice between life and death, depending on the life history of identical cells.

Exactly the same logic applies with even greater force to multicellular organisms. Here the cells are always genetically identical, their fates bound together more tightly than in a loose colony or algal bloom. Even in a simple sphere, differentiation is virtually inevitable: there is a difference between the inside and the outside of the sphere in terms of the availability of nutrients, oxygen and carbon dioxide, exposure to the sun, or threat from predators. The cells couldn't be the same, even if they 'wanted' to be. The simplest adaptations soon begin to pay. At some stages of their development, for example, many algae possess whip-like flagellae that propel them around. In a spherical colony it pays to have such cells with flagellae on the outside, since their combined motion moves the whole colony, while spores (a different developmental stage of genetically identical cells) are protected within. Such a simple division of labour must have given the first primitive colonies a big advantage over single cells. The advantage of large numbers and specialisation is comparable to the first agricultural societies, where for the first time food was plentiful enough to support larger populations, enabling dedication to specialised tasks, such as warfare, farming, metal-working or law-giving. Unsurprisingly, agricultural societies soon displaced small tribes of hunter-gatherers, where comparable specialisation was next to impossible.

Even the simplest colonies already betray a fundamental difference between two types of cell: the germ-line and the 'soma' (the body). This distinction was first noted by the august German evolutionist August Weismann (who we already met in Chapter 5), probably the most influential and insightful nineteenth-century Darwinian after Darwin. Weismann claimed that only the germ-line is immortal, passing genes down from one generation to the next, while the cells of the soma are disposable, mere aids to the immortal germ-line. The idea was discredited for half a century by French Nobel laureate

Alexis Carrel, who was later discredited himself – shamed for fabricating his data. Weismann was right all along. His distinction ultimately explains the death of all multicellular organisms. By its very nature, specialisation means that only some of the cells in a body can be germ-cells; the rest must play a supporting role, their only benefit being the vicarious profit of their shared genes passing down the germ-line. Once somatic cells have 'accepted' their subsidiary role, the timing of their death also becomes subservient to the needs of the germ-line.

The difference between a colony and true multicellularity is best seen in terms of commitment to differentiation. Algae like *Volvox* benefit from community living, but also 'opt out' and live as single cells. Retaining the possibility of independence curtails the degree of specialisation that can be attained. Plainly cells as specialised as neurons could not survive in the wild. True multicellular life can only be achieved by cells 'prepared' to subsume themselves entirely to the cause. Their commitment must be policed, and any attempted reversions to independence are punishable by death. Nothing else works. Just think about the devastation caused by cancer, even today, after a billion years of multicellular living, to appreciate the impossibility of multicellular life when cells do their own thing. Only death makes multicellular life possible. And, of course, without death there could be no evolution; without differential survival, natural selection comes to nothing.

Even in the first multicellular organisms, threatening cells with a death penalty for transgression required no great evolutionary leap. Recall from Chapter 4 that complex 'eukaryotic' cells were formed from a merger between two types of cell: the host cell and the bacteria that later evolved into mitochondria, the tiny powerhouses that now generate energy. The free-living ancestors of mitochondria were a group of bacteria that, like cyanobacteria, possessed the caspase enzymes needed to slice up cells from within. Where they got them from is beside the point (it is possible they were transferred from cyanobacteria, or vice versa, or that both groups inherited them from a distant common ancestor). The point is that mitochondria bequeathed the first eukaryotic cells with the apparatus of death, in full working order.

Whether the eukaryotes could have evolved so successfully into proper multicellular organisms without inheriting caspase enzymes from bacteria is an interesting question, but with caspases there was no stopping them. True multicellularity evolved independently no less than five times in the eukaryotes – in the red algae, green algae, plants, animals and fungi.[3] In their organization these disparate forms of life have little in common, but all of them police their cells and punish transgressions with death using remarkably similar caspase enzymes. Interestingly, in almost all cases the mitochondria are still the principal brokers of death, the hubs inside cells that integrate conflicting signals, eliminate noise, and trigger the death apparatus when necessary. Thus while cell death is necessary for any form of multicellular life, there was little call for evolutionary novelty. The machinery needed was imported into the first eukaryotic cells by way of the mitochondria, and remains broadly the same, if subtly elaborated, even today.

But there is a great difference between the death of cells and the death of whole organisms. Cell death plays an important role in the ageing and death of multicellular organisms, and yet there is no law stipulating that *all* bodily cells should die, or that other equally disposable cells, should not replace them. Some animals, such as the freshwater anemone *Hydra*, are essentially immortal – cells die and are replaced, but the organism as a whole shows no sign of ageing. There is a long-term balance between cell life and cell death. It's like a flowing stream: one can't step into the same stream twice, because the water is forever rushing on and being replaced; but the stream's contours, its volume and shape, remain unchanging. To anyone other than a Greek philosopher, it is the same stream. Likewise the organism, its cells turn over like water, but the organism as a whole is unchanging. I am me, even if my cells change.

This could hardly be any other way. If the balance between cell life and cell death changed, the organism would be no more stable than a stream in spate or draught. Adjust the 'death' settings, to make cell death less likely, and the outcome is cancerous, unstoppable growth. But make cell death too likely and the outcome is a withering away. Cancer and degeneration are two sides of the same coin: both undermine multicellular life. But the simple *Hydra* can maintain its balance forever; and human beings might remain the same weight and build for decades, despite turning over billions of cells a day. Only when we grow old is this balance lost, and then, curiously, we suffer from both sides of the coin at once. Cancer and degenerative diseases are both linked inextricably with old age. So why do organisms grow old and die?

The most popular idea, dating back to Weismann in the 1880s, is wrong, as he himself was quick to recognise. Weismann originally proposed that ageing and death rid populations of old worn-out individuals, replacing them with racy new models replete with a new set of genes remixed by sex. The idea invests death with some sort of nobility and symmetry, in service of a greater cause, even if it can hardly aspire to the grandeur of a religious purpose. In this view the death of an individual benefits the species, just as the death of some cells benefits the organism. But the argument is circular, as Weismann's critics pointed out: old individuals are only 'worn out' if they age in the first place, so Weismann presupposed exactly what he was trying to explain. The question remained, what makes individuals 'wear out' with age, even if death does benefit populations? What's to stop the cheats, individuals who escape death like cancer cells, leaving behind more and more offspring, each endowed with the same selfish genes? What's to stop a cancer on society?

The Darwinian answer was first laid out by Peter Medawar, in his famous inaugural lecture at University College London, in 1953. His answer was that there is a statistical likelihood of death regardless of ageing – of being hit by a bus, or a stone falling from the sky, or being eaten by a tiger, or consumed by disease. Even if you are immortal you are unlikely to live forever. Individuals who concentrate their reproductive resources in the earlier part of their life are therefore statistically more likely to have more offspring than individuals who count on an unhurried schedule, reproducing, say, once every 500 years or so, and regrettably losing their head after only 450. Cram in more sex earlier on and you'll probably leave behind more offspring, who inherit your 'sex-early' genes, than your laid-back cousins. And therein lies the problem.

Each species, according to Medawar, has a statistically probable lifespan, depending on the size of individuals, their metabolic rate, their natural predators, physical attributes like wings, and so on. If that statistical lifespan is, say, twenty years, then individuals who complete their reproductive cycle within that period will normally leave behind more offspring than those who don't. Genes that 'play the odds' will do better than those that don't. Eventually, Medawar concluded, genes that happen to cause heart disease after we are statistically dead will accumulate in the genome. In humans, natural selection cannot eliminate a gene that causes Alzheimer's disease at the age of 150 if nobody lives that long. In past times, genes that caused Alzheimer's at seventy also slipped the net as so few people survived beyond their biblical three score years and ten. Old age, then, for Medawar, is the decline caused by genes that go on operating well after we should be dead – the operation of hundreds, if not thousands, of genes that *are* effectively dead, beyond the reach of selection. Only humans suffer from the misery of Tithonus, for only humans have artificially prolonged our lives by eliminating many of the statistical causes of death, such as predators and many lethal infectious diseases. We have disinterred a graveyard of genes, and they pursue us to our graves.

Medawar's ideas were refined independently by the great American evolutionist George C. Williams, who proposed an idea with one of the worst appellations, surely, in all of science: antagonistic pleiotropy. To me, the term evokes a marine dinosaur goaded into some feeding frenzy. In fact it refers to genes with several effects, some of which are better than others, and indeed some of which are downright harmful. The classic and desperately unfortunate example is Huntington's chorea, a relentless mental and physical degenerative disease, beginning with mild twitches and stumbling, early in middle age, and eventually stripping away the ability to walk, talk, think and reason. This 'lurching madness' is caused by a defect in a single gene, which does not manifest itself until after reproductive maturity. Some tentative evidence suggests that people who go on to develop Huntington's chorea are more likely to be sexually successful earlier in life, although why this should be the case is unknown; and the magnitude of the effect is infinitesimal. But the point is that a gene that engenders even a minuscule degree of sexual success is selected for and remains in the genome, even though it causes the most dreadful degeneration later on.

Quite how many genes are associated with disease late in life is uncertain, but the idea is simple enough and has an explanatory appeal. It's easy to imagine a gene that induces accumulation of iron being beneficial early in life, for example, building up the blood pigment haemoglobin, but detrimental later on, as iron overload causes heart failure. Certainly there's no evolutionary idea more consonant with modern medicine. In popular parlance, there's a gene for everything from homosexuality to Alzheimer's. Yes, this is a figure of speech that helps sell newspapers. But it runs deeper than that. The idea that particular genetic variants predispose to certain conditions permeates the whole of medical research. To give a single well-known example, there are three common variants of the gene *ApoE*, known as *ApoE2*, *E3* and *E4*. About 20 per cent of the population in western Europe has the *ApoE4* variant, and those who do, and who know it, would doubtless wish otherwise: it is associated with a statistically higher risk of Alzheimer's disease, as well as cardiovascular disease and stroke. If you have two copies of the *ApoE4* variant, you'd better eat carefully and get down the gym regularly if you want to offset your genetic predisposition.[4]

What *ApoE4* is actually 'good for' is unknown, but the fact that the variant is so common implies that it might be good for something earlier in life, offsetting its disadvantages later on. But this is just one example of hundreds, if not thousands. Medical research seeks out such genetic variants and attempts to offset their baleful influence through novel (and typically expensive) drugs that target them. Unlike Huntington's chorea, most age-related diseases are a dense tapestry of genetic and environmental factors. In general, multiple genes contribute to the pathology. In cardiovascular disease, for example, diverse genetic variants predispose people to high blood pressure, fast blood clotting, obesity, high cholesterol or laziness. If a propensity to high blood pressure is combined with a salty fat-laden diet, a love of beer and cigarettes, and a preference for television over exercise, we hardly need an insurance company to determine our risk. In general, though, estimating the risk of disease is a thankless task and our understanding of genetic predisposition is still very much in its infancy. Even when added up, the total genetic contribution to age-related diseases is usually less than 50 per cent. Plain old age is almost always the single greatest risk factor – only a few unfortunates succumb to cancer or suffer a stroke in their twenties or thirties.

By and large, then, the modern medical conception of age-related diseases corresponds closely to Medawar's evolutionary picture of late-acting genes. Many hundreds of genes contribute to our predisposition to disease, and each of us has our own spectrum of risk, our own particular graveyard of genes, the effects of which can be exacerbated or ameliorated by our lifestyle, or by other genes. But there are two serious problems with this view of ageing.

The first is implicit in my choice of words here: I am talking about diseases, the *symptoms* of ageing, not the underlying *cause* of ageing. Such genes are associated with particular diseases, but few of them seem to cause ageing itself. It is possible to live to 120, without ever suffering from disease, but nonetheless growing old and dying. For the rest of us, the negative effects of aberrant genes are unmasked by old age: we are not troubled by them when young, only when old. There's a tendency in medicine to view age-related diseases as pathological (and therefore 'treatable'), but to see old age as a 'state', rather than a disease, and so inherently 'untreatable'. There's an understandable reluctance to stigmatise the old as 'diseased'. While this view is not helpful, in that it tries to disengage ageing from age-related diseases, the distinction makes my point about Medawar nicely. He explained the role of genes in age-related diseases, but not the underlying cause of ageing.

The force of the distinction came as a shock in the decade after 1988, when David Friedman and Tom Johnson, at the University of California, Ivine, reported the first life-extending mutation in nematode worms. Dubbed *age-1*, mutations in the gene doubled normal lifespan from around 22 to 46 days. Over the following years, scores of similar mutations were reported in nematodes, as well as other diverse forms of life, from yeast to fruit flies and mice. For a period, the field felt a bit like particle physics in its 1970s heyday, with a veritable zoo of miscellaneous life-extending mutations catalogued away. Gradually a pattern emerged. Almost all the mutated genes encoded proteins in the same biochemical pathway, whether in yeast, flies or mice. In other words, there is an exceptionally conserved mechanism, applying to fungi and mammals alike, which controls lifespan. Mutations in this pathway not only prolong lifespan, but in so doing can also postpone, even evade, the diseases of old age. Unlike poor Tithonus, doubling lifespan more than doubles health span.

The link between disease and lifespan did not come as a surprise. After all, almost all mammals suffer from a similar spectrum of age-related diseases, including diabetes, stroke, heart disease, blindness, dementia, and so on. A rat, however, succumbs to cancer at the age of three years or so, when it is getting old, whereas humans begin to suffer from the same diseases by the age of sixty or seventy. Plainly, even 'genetic' diseases are linked to old age rather than chronological time. What came as a real surprise with the lifespan mutations was the flexibility of the whole system. A single mutation in just one gene can double lifespan and at once put 'on hold' the diseases of old age.

The significance of these findings to ourselves can't be overstressed. All the diseases of old age, from cancer to heart disease to Alzheimer's disease, can in principle be delayed, even avoided, by simple permutations of a single pathway. It's a shocking conclusion, yet it's staring us in the face: it should prove easier to 'cure' ageing and all age-related diseases with a single panacea than it will ever be to cure any one age-related disease like Alzheimer's in people who are otherwise 'old'. This is the second reason I think Medawar's explanation for ageing is wrong. We are not doomed to our own particular graveyard of genes: we can bypass the gene cemetery altogether if we avoid ageing in the first place. Age-related diseases depend on *biological age*, not chronological time. Cure ageing, and we cure the diseases of old age – all of them. And the overriding lesson from these genetic studies is that ageing is curable.

The existence of a biochemical pathway controlling lifespan raises a few evolutionary questions. The first implication, which is erroneous, is that lifespan is written directly in the genes: that ageing and death are programmed, presumably to benefit the species as a whole, as originally proposed by Weismann. But if a single mutation can double lifespan, why don't we see more cheats, more animals that 'opt out' of the system for their own benefit? It would be simple. If animals don't cheat, then there must be a penalty for cheating, one that is sufficiently swingeing to overcome the advantages of a longer life. And if that's the case, we may prefer to keep our diseases.

There is a disadvantage. Sex again. If we want to prolong our lives and avoid disease, it would be wise to read the small print in our contract with death. It's a curious thing that mutations in all the life-extending genes, called *gerontogenes*, act to prolong rather than shorten lifespan. The default position is always shorter lifespan. This makes sense when we consider the nature of the biochemical pathway controlled by gerontogenes. It's not about ageing at all, but sexual maturation. The development of an animal to sexual maturity requires plenty of resources and energy, and if these aren't available it's better to forestall development and wait until they are. This means that environmental plenitude must be monitored and converted into a biochemical currency that speaks directly to cells: 'There's plenty of food, now's a good time to prepare for reproduction. Get ready for sex!'

The biochemical signal that betokens plenitude is the hormone insulin, along with a large family of related hormones that act over longer periods (weeks or months), most notably insulin-related growth factor (IGF). Their names don't concern us – in nematode worms alone there are thirty-nine hormones related to insulin. The point is that, when food is plentiful, the insulin hormones leap into action, orchestrating a range of developmental changes, gearing up for sex. If food is unavailable, these pathways fall silent and sexual development is postponed. But silence doesn't mean that nothing happens. On the contrary, the absence of a signal is detected by other sensors, which in effect put life on hold. Wait, they say, for better conditions, and then try for sex again. In the meantime the body is preserved in a pristine state for as long as possible.

The idea of a trade-off between sex and longevity was laid out by the British gerontologist Tom Kirkwood back in the mid-1970s, long before any gerontogenes had been discovered. Kirkwood pictured exactly such a 'choice', on the economic grounds that energy is limited and everything has a cost. The energetic cost of bodily maintenance must be subtracted from the energetic cost of sex, and organisms that try to do both simultaneously will fare less well than organisms that apportion their resources. The most extreme examples are animals that reproduce just once and don't rear their offspring at all, like Pacific salmon. Their catastrophic demise is then explained not as a programme for death so much as a total investment of resources in the business of life – reproduction.[5] They fall to pieces in days because they allocate 100

per cent of their resources to sex, and withdraw all funding from bodily maintenance. Animals that reproduce more than once, on different occasions, must
allocate fewer resources to sex and more to maintenance; and animals that
invest heavily in raising their offspring over years, like us, adjust this balance
even further. In all cases, though, there is a choice, and in animals that choice
is normally controlled by the insulin hormones.

Mutations in the gerontogenes simulate silence. They disable the signal of
plenty, and instead rouse the genes concerned with bodily maintenance. Even
when food is abundant, the mutant gerontogenes fail to respond. In the first
of several ironies, they resist the sirens' beckoning of insulin. The irony is
that insulin resistance in humans doesn't confer longevity but adult-onset diabetes. The problem is that overeating, coupled with a physiological determi-
nation to hoard scant resources for better times, leads to weight gain, diabetes
and earlier mortality. A second irony: the penalty for prolonging life, defer-
ring sex, remains resolutely in place. It's expressed as infertility. So it's no
fluke that diabetes is linked with infertility. Diabetes and infertility are caused
by the same hormonal swing. Disabling insulin prolongs life only if we're
hungry for much the time, and at the potential cost of not having any
children.

Of course we've known all this for decades – a third irony. We may not
enjoy the fact much, but we've recognised since the 1920s that going moder-
ately hungry prolongs life. It's called calorie restriction. Rats fed a balanced
diet, but with about 40 per cent fewer calories than normal, live half as long
again as their well-nourished siblings, and suffer fewer diseases of old age. As
before, age-related diseases are deferred indefinitely, and are less likely to
strike at all. Whether such calorie restriction exerts the same effects in humans
as rats is uncertain, but the signs are that it does, if to a less dramatic extent:
biochemical studies suggest that broadly similar changes take place in us too.
But despite the fact that we've known about the effects of calorie restriction
for decades, we know remarkably little about how or why it works; or even
whether it really does in humans.

One reason for this is that a proper study of human lifespan would take
decades, which saps the enthusiasm even of the most diligent.[6] Another is the
long-standing perception that living longer means living slower and more

boring lives. That, as it happens, is not true, and gives cause for hope. Calorie restriction improves the efficiency of energy use without lowering overall energy levels; quite the reverse, in fact, it tends to raise them. But the main reason we know so little is that the biochemistry underpinning calorie restriction is a horrendous tangle of feedback loops, parallel circuits and redundancy that shifts kaleidoscopically from tissue to tissue and species to species, and resists unpicking. The significance of the gerontogenes is that they demonstrate that a few trifling changes in this complex network can make a big difference. Not surprisingly, that knowledge had a galvanising effect on researchers.

Calorie restriction is assumed to exert its effects at least in part through the pathways controlled by the gerontogenes. It is a switch: sex or longevity. One problem with calorie restriction is that it flips the switch completely, so there's little possibility of having sex as well as living a long time. But for the gerontogenes, this is not always the case. Some mutations in gerontogenes suppress sexual maturation (the original *age-1* mutation by 75 per cent, for example) but not all of them do so. A few gerontogene mutations turned out to prolong lifespan and health span, with little suppression of sexuality, a mild deferral rather than abolition. Others block sexual development in young animals, but have no obvious negative effects in older adults. Again, the details don't concern us here; the point is that it is possible, with finesse, to disentangle sex from longevity, to activate genes responsible for longevity without dismembering sexuality.

Two gerontogenes that seem to play a central role in calorie restriction have risen to prominence over the last few years, called SIRT-1 and TOR. Both are present almost universally from yeast to mammals, and both exert their effects on lifespan by activating whole suites of proteins. Both are sensitive to the presence or absence of nutrients or growth factors such as the insulin family, and spring into action under reciprocal conditions.[7] TOR is believed to control the sexual side of the switch, by stimulating cell growth and proliferation. It works by switching on other proteins, which between them stimulate protein synthesis and cell growth, while blocking the breakdown and turnover of cellular components. SIRT-1 meanwhile opposes much of this, while mounting a 'stress response' that fortifies the cell. Typically for biology, their activities overlap, but don't counterbalance each other precisely. Between them, though, SIRT-1 and TOR act as the central 'hubs' responsible for coordinating many of the benefits of calorie restriction.

SIRT-1 and TOR sprang to prominence in part because they are definitely important, and in part because we already know how to target them pharmacologically, and that, given the rewards at stake, has fuelled a lively scientific controversy. According to Leonard Guarente, at the Massachusetts Institute of Technology, and his former postdoctoral researcher David Sinclair, now at Harvard, SIRT-1 is responsible for most of the effects of calorie restriction in mammals and can be activated by a small molecule found in red wine known as resveratrol. A string of high-profile publications, beginning with a *Nature* paper in 2003, showed that resveratrol can prolong the lifespan of yeast, worms and flies. Public interest exploded in November 2006, when Sinclair and colleagues published a seminal *Nature* paper, showing that resveratrol reduces the risk of death of obese mice by a third, a finding that was reported on the front page of the *New York Times*, generating a storm of publicity. If it can do that for obese mice, fellow overweight mammals, surely it could work wonders for people too. The well-known health benefits of drinking red wine only added grist to the mill, although the amount of resveratrol in a glass of red wine is barely 0.3 per cent of each dose given to the mice.

Ironically, two former doctoral students in Guarente's lab, Brian Kennedy and Matt Kaeberlein, both of whom are now at the University of Washington, Seattle, have challenged this neat idea recently. Having themselves pioneered some of the early work on SIRT-1, they were troubled by a number of exceptions to their predictions.

In place of SIRT-1, Kaeberlein and Kennedy champion TOR, whose effects, they say, are more pervasive and consistent across species. Because TOR and SIRT-1 have overlapping, rather than precisely opposing properties, they may be right. In particular, blocking TOR represses immune and inflammatory activity, which could be beneficial, because many age-related diseases have a persistent inflammatory component. There's another irony here because TOR actually stands for *target of rapamycin* and was discovered in the context of transplant medicine. Rapamycin is one of the most successful transplant immunosuppressants on the market and has been used for over a decade. It's unusual among immunosuppressants, in that it doesn't make its recipients prone to cancer or bone loss; but few researchers have been persuaded by Mikhail Blagosklonny's strong advocacy of rapamycin as an ideal anti-ageing pill. It will certainly be interesting to see whether transplant recipients receiving rapamycin suffer fewer diseases of old age.

But there is a deeper problem with both resveratrol and rapamycin as anti-ageing medicines, and that is the broad spectrum of their attack. Both coordinate the activation or inactivation of scores or even hundreds of proteins and genes. To some extent this might be necessary, but some parts of this large suite of changes may be unhelpful, or only strictly necessary in the context of short-term food deprivation or stress, which is, after all, the context in which they evolved. So we've seen, for example, that activation of SIRT-1 or inhibition of TOR can potentially induce insulin resistance, diabetes, infertility and immunosuppression. A more targeted approach is preferable, in that fewer trade-offs are likely to be involved.[8] We know that this should be possible because animals that prolong their lifespan by selection over generations in the wild don't suffer any of these drawbacks. The question is, which of the rabble of genes roused by SIRT-1 and TOR are responsible for prolonging life and suppressing disease? Exactly which of the physical changes that take place within cells freeze the passage of time? And can we target them directly?

The answer is not yet known for sure, and, as is so often the way, there seem as many answers as there are researchers. Some emphasise a protective 'stress response', others the up-regulation of detoxifying enzymes, yet others an enhanced waste-disposal system. All may well be important in some circumstances, but their significance seems to vary between species. The single change that appears to be consistent, from fungi to animals including ourselves, concerns those energy-generating powerhouses of the cell, the mitochondria. Calorie restriction almost always induces more mitochondria, which have membranes resistant to damage, and which leak fewer reactive 'free radical' by-products during respiration. These changes are not only consistent, but they mesh beautifully with half a century of research on free radicals in ageing.

✦

The idea that free radicals might cause ageing has its roots in the 1950s, when Denham Harman, who had a background in free-radical chemistry in the oil industry, proposed that these reactive fragments of oxygen or nitrogen (which have lost or gained an electron) may also attack critical biological molecules like DNA and proteins. Harman argued they could ultimately lay waste to cells and drive the process of ageing.

Much has changed in the half-century since Harman's original conception, and it is now fair to say that, as originally stated, the theory is wrong. A subtler version, though, is likely to be right.

There are two things that Harman did not, and could not, know. One is that free radicals aren't merely reactive, but are used by cells to optimise respiration and as a danger signal. They work in much the same way that smoke activates a fire alarm. Rather than attacking proteins and DNA at random, free radicals activate or disable a few key signalling proteins (including TOR itself), which in turn regulate the activity of hundreds of proteins and genes. We now know that free-radical signals are central to cell physiology, so we can begin to see why antioxidants (which mop up free radicals) do as much harm as good. Many still think, following Harman's original prediction, that antioxidants should slow ageing and protect against disease. Repeated clinical studies have proved they don't work. The reason is that antioxidants interfere with free-radical signals. Quashing free-radical signals is equivalent to switching off the fire alarm. To stop this from happening, antioxidant levels in the blood are strictly controlled within tight limits. Large doses of antioxidants are simply excreted or not absorbed in the first place. Antioxidant levels in the body remain roughly constant, keeping the alarm poised for action.

The second factor that Harman didn't know about (because its discovery still lay twenty-five years in the future) was programmed cell death. In most cells, programmed death is still coordinated by the mitochondria, which delivered the entire system into eukaryotic cells 2 billion years ago. One of the principal signals directing cells to take their own lives is an enhanced leak of free radicals from mitochondria. In response to this free-radical signal, the cell trips its death apparatus and silently removes itself from the frame, all traces of its former existence gone. Far from a telltale build-up of molecular debris, envisioned by Harman, the silent machinery of death continuously removes the evidence, with the ruthless efficiency of the KGB. Thus, two of the key predictions of Harman's theory – that molecular damage accumulates to calamitous levels with age and that antioxidants should slow down this accumulation and so prolong life – are simply wrong.

But there are several reasons to think that a subtler version of the theory is broadly correct, even though many details still need to be teased out. First and foremost is the fact that lifespan varies with free-radical leak in virtually all species.[9] The faster the leak of free radicals, the shorter the lifespan. By and large, the rate that free radicals leak depends on the metabolic rate, which is to say, the rate at which cells consume oxygen. Small animals have fast metabolic rates, their cells guzzling up oxygen as fast as they can, their hearts fluttering at hundreds of beats a minute even when at rest. With such fast respiration, free-radical leak is high, and lifespan is fleeting. Larger animals, in contrast, have a slower metabolic rate, manifesting as a ponderous heart beat and a trickling free-radical leak. They live longer.

The exceptions here really do prove the rule. Many birds, for example, live far longer than they 'ought to' on the basis of their metabolic rate. A pigeon, for example, lives for around thirty-five years, a remarkable ten times longer than a rat, despite the fact that pigeons and rats are a similar size, and have a similar metabolic rate. In a groundbreaking series of experiments through the 1990s, the Spanish physiologist Gustavo Barja, at the Complutense University in Madrid, showed that these differences could be accounted for largely in terms of free-radical leak. In relation to their oxygen consumption, birds leak nearly ten times *fewer* free radicals than equivalent mammals. Much the same is true of bats, which also live disproportionately long lives. And like birds, bat mitochondria leak far fewer free radicals. Why this should be so is not certain; in earlier books, I have argued that the reason relates to the power of flight. But whatever the reason may be, the unassailable fact is that low free-radical leak equates to long life, whatever the metabolic rate.

It's not just lifespan that varies with free-radical leak, health span does too. We've noted that the onset of age-related diseases doesn't depend on the passage of time, but on biological age. Rats and humans suffer from the same diseases, but in rats they start within a couple of years, while in humans they take many decades. Some degenerative conditions are caused by exactly the same mutations in both rats and humans, yet we always see the same time difference. The aberrant genes that Medawar linked with ageing, and which are so central to medical research, are unmasked in old animals by something about the state of their aged cells. Alan Wright and colleagues at the University of Edinburgh have shown that this 'something' is linked to the rate of free-radical leak. If free-radical leak is fast, degenerative diseases set in quickly; if it's slow, they're postponed or even abrogated altogether. Birds, for example, suffer from few of the age-related diseases common to most mammals (bats excepted, again). A reasonable hypothesis is that free radical leak ultimately alters the state of cells, making them 'old', and that this altered state unmasks the negative effects of late-acting genes.

How do free radicals alter the state of cells during ageing? Almost certainly through unintended effects on signalling. The use of free radical signals optimises our health when we're young, but has detrimental effects when we're older (as in G. C. Williams's theory of antagonistic pleiotropy). As the mitochondrial populations of cells begin to wear out, free-radical leak starts to creep up, if only slightly, until it passes a threshold that activates the fire alarm – continuously. Hundreds of genes are switched on in a vain attempt to restore normality, giving rise to the chronic, albeit mild, inflammation that's characteristic of many of the diseases of old age.[10] This persistent mild inflammation alters the properties of many other proteins and genes, placing the cell under even greater stress. It is this pro-inflammatory state, I suspect, that unmasks the detrimental effects of late-acting genes like *ApoE4*.

There are only two ways out from here. Either cells can cope with a chronically stressed state or they can't. Different cell types vary in their ability to cope, depending largely on their 'job'. The best example I know of comes from the work of pioneering pharmacologist Salvador Moncada, at University College London. Moncada has shown that neurons and their support cells, known as astrocytes, have diametrically opposed fates. Neurons depend on their mitochondria. If they can't generate enough energy to meet their needs with mitochondria, the cell death apparatus is triggered and the neuron is silently eliminated. The brain shrinks, by as much as a quarter by the time the early symptoms of Alzheimer's disease become apparent. In contrast, astrocytes can survive quite happily without mitochondria. They switch to alternative energy sources (the *glycolytic switch*) and become virtually immune to programmed cell death. These two opposing outcomes explain why both degeneration and cancer go hand in hand with ageing. If they can't switch over to alternative energy, cells die, causing degenerative diseases as tissues and organs shrink, heaping ever more responsibility on the few remaining cells. On the other hand, cells that can switch over do so and become virtually immune to cell death. Goaded by the unceasing inflammatory conditions, they proliferate, swiftly accumulating mutations that free them from the normal constraints of the cell cycle. They transform into cancer cells. It's no fluke that neurons rarely, if ever, form tumours, whereas astrocyctes are relatively common perpetrators.[11]

From this perspective, we can see why calorie restriction protects against age-related disease, as well as ageing, at least if started early enough in life (before the mitochondria wear out: middle age is fine). By lowering free-radical leak, bolstering mitochondrial membranes against damage, and boosting the number of mitochondria, calorie restriction effectively 'resets' the clock of life back to 'youth'. In so doing, it switches off hundreds of inflammatory genes, returning genes to their youthful chemical environment, while fortifying cells against programmed cell death. The combination suppresses both cancer and degenerative diseases and slows the rate of ageing. It's likely that, in practice, various other factors are involved (such as the direct immunosuppressive effects of inhibiting TOR), but in principle most benefits of calorie restriction can be explained simply by a reduction in free-radical leak. It makes us more like birds.

There's an intriguing piece of evidence that this really is how it all works. In 1998, Masashi Tanaka and his colleagues, then at the Gifu International Institute of Biotechnology in Japan, examined the fate of people with a common variant in their mitochondrial DNA (common in Japan at least, if not, unfortunately, elsewhere in the world). The variant alters a single DNA letter. The effect of the change is a tiny reduction in free-radical leak, barely detectable at any one moment, but sustained over a lifetime. The consequences, however, are immense. Tanaka and his colleagues sequenced the mitochondrial DNA of several hundred patients who arrived consecutively in hospital, and found that up to the age of around fifty there was no difference in the proportion of the two types, the variant of interest and 'normal'. But then, after the age of fifty, a gap started to open up between the two groups. By the age of eighty, people with the variant were half as likely to arrive in hospital for any reason at all. And the reason they didn't come to hospital wasn't because they were dead or something. Tanaka found that the Japanese with the variant are twice as likely to live to 100. This implies that people with the variant are half as likely to suffer from any age-related disease. Let me say that again, as I don't know any other fact as shocking in all of medicine: a tiny change in the mitochondria halves our risk of being hospitalised for any age-related disease and doubles our prospects of living to 100. If we are serious about tackling the distressing and cripplingly expensive health problems of old age on our greying planet, this, surely, is the place to start. Shout it from the rooftops!

I don't want to downplay the scientific challenge of the times ahead or denigrate the efforts of researchers who have focused their whole careers on the minutiae of particular diseases of old age. Without their heroic success in elucidating the genetics and the biochemistry of disease, no broader synthesis is possible. There is, nonetheless, a danger that medical researchers are either unaware of evolutionary thinking or uninterested. If nothing in biology makes sense except in the light of evolution, as argued by the evolutionary thinker Theodore Dhobzhansky, medicine is even worse: the modern view of disease holds no meaning whatsoever. We know the price of everything and the value of nothing. The stoical generation of my grandparents used to console themselves by saying that these things are sent to try us, but as this fatalism has faded from the bedside, diseases now just happen, blighting lives on a scale that puts the four horsemen of the apocalypse to shame. It's now a 'fight' against cancer or Alzheimer's, a fight we know we will one day lose.

But death and disease are not random. They do hold meaning, and we can use that meaning to cure ourselves. Death evolved. Ageing evolved. They evolved for pragmatic reasons. In the broadest of terms, ageing is flexible, an evolutionary variable that is set against various other factors, like sexual maturation, in the ledger book of life. There are penalties for tampering with these parameters, but the penalties vary and in a few cases at least can be trivial. In principle, small adjustments to particular pathways should enable us to lead longer, healthier lives. Let me put that more strongly. Evolutionary theory suggests that we can eradicate the diseases of old age with a single well-judged panacea. The anti-ageing pill is not a myth.

But I suspect that a 'cure' for Alzheimer's disease is a myth. In fact, medical researchers don't much like the word 'cure', preferring instead more careful terms like 'ameliorate', 'palliate' or 'delay'. I doubt that we will ever cure Alzheimer's in people who are otherwise 'old', because we're ignoring the terms of the evolutionary deal. It's like trying to patch up a few of the cracks in a leaking dam with putty and hoping to stay the flow. Much the same applies to stroke, heart disease, many forms of cancer, and so on. We have uncovered an impressive amount of detail. We know what is going on protein by protein, gene by gene, yet we are missing the wood for the trees. Such diseases happen to old bodies, they are a product of an old internal environment, and if we intervene early enough in life, we can reset that environment to 'young', or at least 'younger'. It won't be straightforward; there are too many contractual details for that, too many trade-offs. But if we spend a fraction of the time and effort dedicated to medical research on the underlying mechanisms of ageing, I would be amazed if we didn't have an answer within the next two decades. An answer that cures all the diseases of old age at once.

Some people will be concerned about the ethics of extending lifespan, but I suspect that it won't necessarily be a problem. The longevity dividend of calorie restriction, for example, apparently declines with lifespan. While rats nearly double their lifespan, the same extension is not seen in rhesus monkeys. The monkey studies aren't yet complete, but it's likely that in terms of lifespan the benefits will be more modest. The health benefits, though, may be a different matter. The biochemical changes in rhesus monkeys imply that old monkeys suffer fewer of the diseases of old age, even though their lifespan is not greatly prolonged. My hunch is that it will be easier to prolong health span than lifespan. If we can invent an anti-ageing pill that mimics the benefits of calorie restriction, while avoiding the drawbacks, we're likely to see a general improvement in health and many more healthy centenarians, something akin to the lucky Japanese with the mitochondrial variant, but I doubt that we'll see anyone living to 1,000, or even 200. That is a more exacting mission, should we wish to accept it.[12]

The chances are we will never live forever, nor would many wish to. The problem is implicit in the make-up of the first colonies, the distinction between the sex cells and the body. Once cells began to differentiate, the disposable soma became subservient to the germ-line. The more that cells became specialised, the greater the benefits to the body as a whole, and to the germ-line in particular. The most specialised cells of all are the neurons of the human brain. Unlike more mundane cells, neurons are practically irreplaceable, each one wired up with as many as 10,000 synaptic connections, each synapse founded in our own unique experience. Our brains are not replaceable. When our neurons die, there is usually no pool of stem cells to replace them; and if one day we succeed in engineering a pool of neuronal stem cells, we must surely replace our own experience in the bargain. And so the price of immortality is our humanity.

EPILOGUE

In one of the most arresting sequences ever screened on TV, Jacob Bronowski paced through the marshes at Auschwitz, where the ashes of 4 million people had been flushed, including some of his own family, and talked to camera in the way that only he could. Science, he said, does not dehumanise people and turn them into numbers. Auschwitz did that. Not by gas, but by arrogance. By dogma. By ignorance. It happens, he said, when people aspire to the knowledge of gods, and have no test in reality.

Science, in contrast, is a very human form of knowledge. Bronowski put it beautifully: 'We are always at the brink of the known; we always feel forward for what is to be hoped. Every judgement in science stands on the edge of error, and is personal. Science is a tribute to what we *can* know although we are fallible.'

This scene, from *The Ascent of Man*, was screened in 1973. The following year, Bronowski died of a heart attack, as humanly fallible as science itself. But his inspiration lives on, and I know of no better testament to the spirit of science. And in this spirit, and in a title that pays oblique homage, this book has walked the brink of the known. It is full of judgements that stand on the edge of error. It is a tribute to what we can know although we are fallible.

But where is this line between fallacy and truth? Some scientists may disagree with details in this book; others will agree. Disagreements happen on the edge of error, and it is easy to fall over the brink. But if the details are shifting or wrong, does that make the larger story wrong too? Is scientific knowledge relative, especially when applied to the deep past? Can it then be challenged, as it is every day by those who prefer the comfort of dogma? Or is the science of evolution just one more dogma, refusing to countenance challenge?

The answer, I think, is that evidence can be at once fallible and overwhelming. We can never know the past in all its detail, for our interpretations are always fallible, always open to more than one reading. That's why science can be so controversial. But science has a unique power to settle scores through experiment and observation, through tests in reality, and the countless details give rise to something bigger, just as, with the right distance, innumerable pixels paint a compelling picture. To doubt that life evolved, even if some of the details described in this book may yet prove wrong, is to doubt the convergence of evidence, from molecules to men, from bacteria to planetary systems. It is to doubt the evidence of biology, and its concordance with physics and chemistry, geology and astronomy. It is to doubt the veracity of experiment and observation, to doubt the testing in reality. It is, in the end, to doubt reality.

I think that the picture painted in this book is true. Life most surely evolved, along the lines described here. That is not dogma, but evidence tested in reality and corrected accordingly. Whether this grand picture is compatible with faith in God, I do not know. For some people, intimately acquainted with evolution, it is; for others, it is not. But whatever our beliefs, this richness of understanding should be a cause for marvel and celebration. It is a most wonderful thing to share so much with the life around us on this blue-green marble, floating through the bleak infinity of space. There is more than grandeur in this view of life. There is fallibility and majesty, and the best human eagerness to know.

NOTES

Chapter 1

1. Specifically, this is a redox reaction, in which electrons are transferred from a donor (hydrogen) to an acceptor (oxygen), which wants them a lot more, to form water, a thermodynamically stable end product. All redox reactions involve electron transfer from a donor to an acceptor; and remarkably, all life, from bacteria to man, relies on electron transfers of one sort or another for its energy. As the Hungarian Nobel laureate Albert Szent-György put it, 'Life is nothing but an electron looking for a place to rest.'

2. This statement is not strictly true. The vents emit a faint light (discussed in Chapter 7), which is too dim to be detected by the human eye, yet is strong enough to power photosynthesis in some bacteria. But these contribute little to the abundance of the ecosystem compared with the sulphur bacteria. Incidentally, the irrelevance of heat and light was confirmed by the discovery of cold-seepage sites on the ocean floor, where much the same exuberant fauna exists as in these vents.

3. Other problems include the temperature (some say too hot for organic molecules to survive), the acidity (most black smokers are too acid to support the chemistry that Wächtershäuser proposes, and his own lab syntheses only worked in alkaline conditions), and sulphur (too much, relative to modern biochemistry).

4. There is an interesting question about the long-term consequences of cooling the planetary core. As the mantle cools, seawater will tend to bind to the rocks and remain part of their structure, rather than being driven off by heat and thrust back to the surface through active volcanism. A cooling planet might finally consume its own oceans in this way; such a process might have contributed to the loss of oceans on Mars.

5. There are two domains of simple prokaryotic cells lacking a nucleus: bacteria and archaea. The main tenants of the Lost City are archaea that derive their energy through the production of methane (methanogenesis). Archaea have a very different biochemistry to the complex eukaryotic cells that make up plants and animals. So far, no known pathogens or parasites derive from within the archaea; all are bacteria, which share far more biochemistry with their host cells. Maybe the archaea are just too different. One exception is the partnership between an archaeon and a bacterium that may have given rise to the eukaryotic cell itself 2 billion years ago – see Chapter 4.

6. The chemical name for vinegar is acetic acid, giving the root 'acetyl'; in a thioester, this two-carbon molecule is tethered to a reactive sulphur group. For two decades now, Christian de Duve has been extolling the central importance of acetyl thioesters in early evolution, and his arguments are finally being taken seriously by experimentalists.

7. For those who want the full story, along with more on the strangeness and cosmic importance of chemiosmosis, try my earlier book *Power, Sex, Suicide: Mitochondria and the Meaning of Life*.

Chapter 2

1. With so many new mutations you might be wondering why we don't all undergo a mutational meltdown, and the same question troubles many biologists. The answer, in a word, is 'sex', but the explanation is for Chapter 5.

2. This figure refers to DNA sequence similarity. Other larger genomic changes, such as deletions and chromosome fusions, have also taken

place since the chimps and humans diverged, giving an overall genome similarity closer to 95 per cent. In comparison, genetic differences between human populations are tiny – we are 99.9 per cent genetically identical. Such restricted variation reflects a relatively recent population 'bottleneck', perhaps 150,000 years ago, in which a small population in Africa gave rise to all modern human races, via successive waves of migration from Africa.

3. The T (thymine) in DNA is replaced with a slightly different base, called uracil (U) in RNA. This is one of only two tiny differences in structure between RNA and DNA, the other being the use of a sugar called ribose instead of the deoxyribose in DNA. We'll see later what a big functional difference these two tiny chemical details make.

4. So how does nature avoid the frame-reading problem? Simple: it starts at the start and ends at the end of mRNA. Instead of tRNAs lining up like piglets on the mother's nipples, the process is strikingly mechanical. The mRNA is fed like an audiotape through a ribosome, which works as a tape-reader, reading off each codon in turn, until it reaches the end. Instead of zipping up the whole protein when finished, the protein is extended bit by bit, until finally it is released when the ribosome reaches the end. Several ribosomes can work their way down the same strand of messenger RNA at once, each one building a new protein as it goes.

5. The names don't concern us, but for what it's worth: if the first letter of the code is a C, the amino acid encoded is derived from α-ketoglutarate; if an A, the amino acid derives from oxalocetate; if T, pyruvate. Finally, if the first letter is a G, the amino acid is formed in a single equivalent step from any of a number of simple precursors.

6. It is possible that the transfer of the amino acid to the RNA depends on the sequence of the RNA. Michael Yarus and colleagues at the University of Colorado have shown that small RNA molecules containing multiple anticodon sequences bind the 'correct' amino acid with up to a millionfold greater affinity than other amino acids.

7. In the lab an enzyme is needed too – a DNA polymerase. It's likely that an enzyme would be needed to promote RNA or DNA replication in the vents too; but there's nothing to say that the enzyme should be a

protein. An RNA replicase should do just as well, and this has become a
bit of a holy grail, although its existence seems likely.

8. Our own (eukaryotic) method of DNA replication is taken from the
    archaea, not the bacteria, for reasons that we'll explore in Chapter 4.
9. Watson and Crick observed: 'It is probably impossible to build this
    structure [a double helix] with a ribose sugar in place of the
    deoxyribose, as the extra oxygen atom would make too close a van der
    Waals contact.'

Chapter 3

1. There is about 550 times more oxygen than carbon dioxide in the
    atmosphere, obviously making it much easier to double or treble carbon
    dioxide levels. However, even though atmospheric oxygen levels have
    not changed much at all, rising temperatures lowers the solubility of
    oxygen in water. Already fish populations are being affected by low
    dissolved oxygen levels in the oceans. For example, the eelpout
    populations in the North Sea vary year by year with dissolved oxygen
    concentration – the lower the oxygen, the smaller the population.
2. For more on the role of oxygen in evolution, see my earlier book
    Oxygen: The Molecule that Made the World.
3. If you want to know more I can heartily recommend Oliver Morton's
    book *Eating the Sun*.
4. T. H. Huxley, on reading *The Origin of Species*, exclaimed 'How
    extremely stupid not to have thought of that!'
5. In the electromagnetic spectrum, energy and wavelength are inversely
    related; the lower the wavelength the higher the energy. Chlorophyll
    absorbs light in the visible part of the spectrum, specifically red light.
    The powerful oxidant form of chlorophyll is known as P680, as it
    absorbs light with a wavelength of 680 nanometres. Other forms of
    plant chlorophyll absorb slightly less energetic light, with a wavelength
    of 700 nanometres. Blue and yellow light are not needed for
    photosynthesis at all, and so are reflected back (or transmitted through)
    – and this is why we see plants as green.

6. For those who wonder how biochemistry got its bad name, NADPH stands for the reduced form of nicotinamide adenine dinucleotide. It is a strong 'reductant', which is to say a strong pusher of electrons.

7. Technically speaking, they are not called photosystems in bacteria, but photosynthetic units. However, the bacterial reaction centres prefigure the plant photosystems so exactly in both structure and function that I shall stick with the same term.

8. Porphyria is actually a group of diseases caused by porphyrins accumulating in the skin and organs. Most forms are quite benign, but sometimes the porphyrins which collect are activated by light and cause the most distressing burns. The worst forms of the disease, such as chronic erythropoietic porphyria, are so destructive that the nose and ears can be eaten away, and the gums eroded leaving teeth protruding like fangs, with scar tissue and hair growing on the face. Some biochemists have linked these conditions with folk legends of vampires and werewolves, much to the anger of people with milder forms of the disease, who feel they have enough to contend with without being labelled outcasts. In fact, the worst forms of porphyria are rarely seen these days, as precautions and better treatments have limited the nastiest afflictions. More positively, the caustic properties of light-sensitive porphyrins have been put to good use as a treatment for cancer – photodynamic therapy – in which light is used to activate porphyrins targeted to the tumour.

9. According to John Allen, the two photosystems diverged in an ancestor of the cyanobacteria under different usage; others argue that the two photosystems diverged in totally different lines of bacteria, and were brought together later by some kind of gene fusion, forming a genetic chimera that was itself the ancestor of modern cyanobacteria. Recent evidence favours Allen's view (by suggesting that the photosystems passed from cyanobacteria to other lines rather than the other way around) but at present the genetic evidence is equivocal. Either way, however, the photosystems must have functioned independently to begin with.

10. According to Jim Barber, this is exactly how the oxygen-evolving complex forms today. If the complex is removed from Photosystem II, and the 'empty' photosystem is placed in a solution containing manganese and calcium ions, a series of light flashes can reconstitute the complex. Each flash oxidises one manganese ion, which, when oxidised, binds into place. After five or sex flashes, all the manganese and calcium ions are in place, reconstituting the full oxygen-evolving complex. In other words, given the right protein context, the complex is self-assembling.

Chapter 4

1. By the time you read this, Windows XP may mean no more to you than Windows 286. It will have vanished too, no doubt replaced by a yet more sophisticated (if unstable and virus-prone) system.
2. That doesn't mean to say there *aren't* any bacterial equivalents. For example, the bacterial cytoskeleton is composed of proteins that are plainly related to the eukaryotic equivalents, because their physical structure is so similar that they can be superimposed in space. Even so, the gene sequences have diverged to the point of losing all identity. If judged on gene sequence alone, the cytoskeletal proteins would be considered uniquely eukaryotic.
3. Woese maintains that his ribosomal RNA tree is 'canonical' because the gene for small-subunit ribosomal RNA is not just slowly evolving, but is *never* exchanged by lateral gene transfer: it is only ever inherited vertically, which is to say from mother to daughter cell. However, this is not strictly true; there are known examples of lateral transfer of the ribosomal RNA gene among bacteria such as *Neisseria gonorrhoea*. How common this is over evolutionary time is another question, one that can only be answered through more sophisticated 'consensus' trees using many genes.
4. This is a cellular version of that old philosophical chestnut about identity: would we retain our sense of 'self' if we exchanged all but the bits of brain responsible for our memory? And if our memories were

transplanted into another person, would they assume the donor's persona? A cell, like a person, is surely the sum of all its parts.

5. Evolution, of course, does both, and there is no contradiction; the difference boils down to whether the speed of change is measured in generations or geological aeons. Most mutations are damaging and are eliminated by natural selection, leaving everything the same, unless changes in the environment (like mass extinctions) alter the status quo. Then change can be rapid in terms of geological time; but it is still mediated by the same processes at the level of the genes, and it is still slow in terms of change from one generation to the next. Whether catastrophes or small changes are emphasised depends a great deal on temperament – whether scientists are given to cry revolution!

6. The 'hydrogen hypothesis' of Bill Martin and Miklós Müller postulates that the relationship is between an archaeon, which grows on hydrogen and carbon dioxide, and a bacterium capable of respiring with oxygen or of fermenting to produce hydrogen and carbon dioxide, as circumstances demand. This versatile bacterium can presumably make use of the methane released as a waste product by the archaeon. But I won't discuss this idea any further here, as I did so at some length in an earlier book, *Power, Sex, Suicide: Mitochondria and the Meaning of Life*. The ideas discussed in the next few pages are developed in more detail in that book.

7. Technically, the surface-area-to-volume ratio falls with greater size, because the surface area rises as a square, while volume rises as a cube. Doubling the linear dimensions quadruples the surface area ($2 \times 2 = 4$) but raises the volume eightfold ($2 \times 2 \times 2 = 8$). The outcome is that energetic efficiency tails away as bacteria get bigger – the membrane used for generating energy becomes smaller relative to cell volume.

8. I have argued this case in lectures around the world and have yet to be faced with a 'killer' criticism. The closest is from Cavalier-Smith, who points to a few eukaryotic cells that phagocytose without mitochondria today. I don't think their existence refutes the argument, as the strongest selection pressures are against prokaryotes that respire over their outer membrane. Once a phagocyte has evolved, it is possible to

whittle away at one or another of its parts under various circumstances, a process of reductive evolution common in parasites. It is much easier for a fully evolved phagocyte to lose mitochondria under certain circumstances, like parasitism, than for a prokaryote to evolve into a phagocyte without the aid of mitochondria.

Chapter 5

1. Some say the lady was Mrs Patrick Campbell, England's most celebrated and notorious actress, for whom Shaw later wrote the part of Eliza Doolittle in *Pygmalion*; others say she was the scandalous mother of modern dance, Isadora Duncan. But the story is probably apocryphal.

2. In Uganda, one of the few African countries to turn the tide, the prevalence of HIV has fallen from 14 per cent to 6 per cent over a decade, largely the result of better public information. The message is simple enough in principle, if not in practice: avoid unprotected sex. One study showed that the success of Uganda's 'ABC route' – Abstinence, Be faithful, use Condoms – is largely attributable to the last of these.

3. Behaviour, incidentally, predicted by Richard Dawkins in *The Selfish Gene*, and since borne out beyond even his penetrating vision.

4. Bacteria do not replicate strictly clonally in this sense, because they also acquire DNA by lateral transfer from other sources. In this sense, bacteria are far more flexible than asexual eukaryotes. The difference that it makes for bacteria can be seen in the rapid spread of antiobiotic resistance, which is usually achieved via the lateral transfer of genes.

5. The story is told by Matt Ridley, with trademark flair, in his book *The Red Queen*, first published in 1993.

6. You may object that the immune system evolved to do that. True enough, it did, but in fact the immune system has a weakness that can only be fixed by sex. To work, the immune system must distinguish between 'self' and 'foreign'. If the proteins that define 'self' remain the same generation after generation, then all that a parasite must do to

evade the immune system is to camouflage itself in proteins that look like 'self': ignore all the smoke and mirrors and go for the sitting duck. This is what would happen to any clones that have an immune system. Only sex (or very high mutation rates at critical targets) can change the immune system's conception of 'self' every generation.

7. Not quite true. One reason that the Y chromosome hasn't disappeared altogether is that it contains multiple copies of the same gene. The chromosome apparently bends double, allowing recombination between the genes on the same chromosome. Even such restricted recombination appears to have saved the Y chromosome from oblivion, at least in most mammals. Some, though, like Asiatic mole voles, have lost their Y chromosome altogether. How their males are specified is uncertain, but it's reassuring that men aren't necessarily doomed along with our degenerate chromosome.

8. These two statements say nothing about the identity of the host cell, or the nature of the symbiotic union between cells, two aspects that are particularly controversial. Nor does it matter, in this scenario, whether the host cell has a nucleus, or a cell wall or a phagocytic lifestyle. So while the origin of the eukaryotic cell is controversial in many respects, the ideas outlined here are independent of any one theory.

Chapter 6

1. William Croone was one of the founding members of the Royal Society, and his name lives on in the Croonian Lecture, the Society's premier lecture in biological sciences.

2. Churchill famously wrote: 'History will be kind to me, for I intend to write it.' His magisterial writings were duly awarded the Nobel Prize for Literature in 1953. When did history last win the prize for literature?

3. Perutz and Kendrew first solved the structure of sperm whale myoglobin, which might seem a curious choice. The reason was that myoglobin had been found to crystallise from pools of blood and gore on the decks of whalers (it is present at much higher concentrations in the muscles of diving mammals like whales). This tendency to

crystallise is critical, as some form of crystal, or at least repetitive structure, is needed for crystallography to work at all.

4. Different muscles are composed of a mix of different fibres. Fast-twitch fibres depend on anaerobic respiration for their power, which is fast but inefficient: such fibres contract quickly (with fast myosins) but tire easily. They have little need for rich capillary networks, mitochondria, or myoglobin, the accoutrements of aerobic respiration, and this gives them a predominantly white colour, reflected in white meat. Slow-twitch fibres are found mainly in red meat, and rely on aerobic respiration (with slower myosins). They contract more slowly but fatigue less easily.

5. This is actually a slight simplification: the gene sequences are 80 per cent identical, but the sequence of amino acids in the protein are 95 per cent identical. This is possible because there are several ways of encoding the same amino acid (see Chapter 2). The discrepancy reflects regular mutations in gene sequence, coupled with a strong selection for retaining the original protein sequence. Almost the only changes in gene sequence permitted by selection are those that do not change the identity of the amino acids in the chain. Just another small sign of selection in action.

6. Of course, these changes actually happened the other way around: the processive motors eventually became the thick filaments of muscle. Perhaps this explains why each myosin molecule still has two heads in muscle, even though they don't seem to be usefully coordinated.

7. Bacteria can also move around, using a flagellum that is very different to anything in eukaryotic cells. It is basically a rigid corkscrew, rotated on its axis by a protein motor. The bacterial flagellum is frequently touted as an example of 'irreducible complexity', but the argument has been dealt with so comprehensively elsewhere that I don't intend to tackle it here. If you want to find out more about the bacterial flagellum, do please read *The Flagellum Unspun* by Ken Miller, an eminent biochemist, scourge of the intelligent design movement, and a practising Catholic. He sees no contradiction between the belief that the molecular details of life are all explained by evolution and a belief

in God. Advocates of intelligent design, however, he dismisses as a double failure, 'rejected by science because they do not fit the facts, and having failed religion because they think too little of God'.

8. Specifically the G-proteins, a family of molecular 'switches' involved in cell signalling. The bacterial relatives are a large family of GTPase proteins. The names are not important; suffice to say that the protein ancestors are known.

9. An even less helpful example is bovine spongiform encephalopathy, or BSE, otherwise known as mad-cow disease. BSE is an infectious disease transmitted by prions, which is to say proteins that act as infectious particles. These alter the structure of adjacent proteins. The altered proteins polymerise into long fibrils, in other words, they form a sort of cytoskeleton. Although assumed to be simply pathological, recent work suggests that prion-like proteins may play a role in the formation of long-term memory at synapses in the brain.

Chapter 7

1. One of the few claims to fame of my old school was a boy who went on to cox for Cambridge in the boat race with Oxford. He steered the Cambridge boat straight into a barge, and sank, along with his very despondent crew. He later explained the barge was in his blind spot.

2. 'I could see that, if not actually disgruntled, he was far from being gruntled.'

3. Did you know, for example, that most mammals (primates apart) have no ability to accommodate, which is to say, adjust their focus from long-distance to close-up? It's an added extra.

4. Ammonites fell extinct with the dinosaurs leaving their magnificent spiral shells behind as fossils in Jurassic rocks. My favourite specimen is embedded on a vertiginous and, for an ageing rock climber, tantalisingly unattainable ledge on the sea cliffs at Swanage in Dorset.

5. The final step to evolve a trilobite eye, not included in Figure 7.2, would be the duplication of ready-made facets to form a compound eye. This is not a problem: life is good at duplicating existing parts.

6. My own favourite is the tiny parasitic flatworm *Entobdella soleae*, which has a lens made from mitochondria fused together. Mitochondria are normally the 'powerhouses' of complex cells, generating all the energy we need to live, and they certainly don't have any special optical properties. Indeed, other flatworms have lenses made from clusters of mitochondria that don't even bother to fuse. Apparently a cluster of ordinary cellular components bends light well enough to serve some advantage.

7. The Bell Labs team were actually interested in the commercial production of microlens arrays for use in electronic and optical devices. Rather than trying to fashion these arrays with lasers, the normal but flawed technological approach, the team followed the lessons of biology – the buzzword is biomimetics – and let nature do it for them. Their success was reported in *Science* in 2003.

8. The late Sir Eric Denton, head of the Marine Biology Association lab in Plymouth, told a good variant: 'When you get a good result have a good dinner before you repeat it. Then at least you've had a good dinner.'

9. The sharp-eyed, or forewarned, will have noticed that the 'red' cone absorbs maximally at 564 nanometres, which is not red at all, but in the yellowish-green part of the spectrum. For all its vividness, red is a complete figment of the imagination: we 'see' red when the brain assimilates the information from two different cones: no signal at all from the green cone, and a waning signal from the yellowish-green cone. It just goes to show the power of the imagination. Next time you argue about whether two dissimilar shades of red match, remind your adversary that there's no 'right' answer; she must be wrong.

10. As all paparazzi know, the bigger the lens, the better it sees; and the same applies to eyes. The reverse is obviously true, and sets a lower limit to the size of the lens, somewhere near the size of the individual facets of the insect compound eye. The problem hinges not just on lens size, but also on the wavelength of light – shorter wavelengths give better resolution. Perhaps this explains why insects today, and the early (small) vertebrates, see in the ultraviolet: it gives better resolution in

small eyes. We don't need it because we have a large lens, and so can afford to cut out this more damaging part of the spectrum.

Interestingly, the ability of insects to see in the ultraviolet means that they can perceive patterns and colours that exist in flowers which we see simply as white. That helps account for why there are so many white flowers in the world: to the pollinators themselves, they are richly patterned.

11. Bacterial rhodopsins are common. They have a similar structure to both algal and animal rhodopsins, and their gene sequences are related to the algal rhodopsins. Bacteria use rhodopsins both as light sensors and for a form of photosynthesis.

Chapter 8

1. Clement Freud was the grandson of Sigmund Freud and for a time a Liberal politician. On an official trip to China, he was surprised to find that his junior colleague had been given a grander suite. It was explained that his colleague was the grandson of Winston Churchill. 'It's the only time I've ever been out-grandfathered!' recalled Freud.

2. This is not strictly true. Larger animals produce less heat per pound than small animals, that is, metabolic rate falls with size. The reasons are controversial and I won't go into them here. For those who want a full discussion, see my earlier book *Power, Sex, Suicide: Mitochondria and the Meaning of Life*. Suffice to say that large animals do indeed retain heat better than small animals, even though they generate less heat per pound.

3. With apologies to blues legend Howlin' Wolf. 'Some folk built like this, some folk built like that. But the way I'm built, you shouldn't call me fat. Because I'm built for comfort, I ain't built for speed.'

4. If you have difficulty conceiving of how all these traits could be selected for at once, just look around you. Some people are obviously more athletic than others – a small proportion of individuals are blessed with Olympic stature. You may not wish to accept this as your mission, but a programme breeding athletes with athletes and selecting only the

fittest would almost certainly succeed in producing 'superathletes'. Such experiments have been done in rats, to study diabetes, and 'aerobic capacity' was improved by 350 per cent in just ten generations (lowering the risk of diabetes). They also lived six months longer – a lifespan extension of about 20 per cent.

5. One interesting possibility, argued forcefully by Paul Else and Tony Hulbert, at the University of Wollongong in Australia, relates to the lipid composition of cell membranes. A fast metabolic rate demands the fast passage of materials across membranes, and this generally requires a relatively high proportion of polyunsaturated fatty acids, whose kinked chains produce greater fluidity – the difference between lards and oils. If animals are selected for high aerobic capacity, they tend to have more polyunsaturated fatty acids, and their presence in visceral organs may force a higher resting metabolic rate too. The drawback is that it should be possible to vary the fatty acid composition of membranes in different tissues, and this certainly happens to some extent. So I'm not convinced it quite solves the problem. Nor does it explain why there should be more mitochondria in the visceral organs of hot-blooded animals. This implies that a high metabolic rate has been deliberately selected for in these organs, rather than being an accident of lipid composition.

6. The discovery of a fossil lystrosaur in Antarctica by Edwin Colbert in 1969 helped confirm the then controversial theory of plate tectonics, for lystrosaurs had already been found in southern Africa, China and India. It was easier to believe that Antarctica floated there than that the squat lystrosaurs swam.

7. The essential insight, according to Richard Prum at Yale, is that feathers are basically tubular. Tubes are important from an embryological point of view, because they have several axes: up/down, across, and inside/outside. Such axes generate biochemical gradients, as signalling molecules diffuse down the axes. The gradients in turn activate different genes along the axes, which control embryonic development. Bodies, too, are essentially 'tubes' to embryologists.

8. As an ex-smoker and mountaineer, used to gasping for breath at all altitudes, I can only begin to imagine the rush that birds would get from smoking – the impact of continuous-infusion, high-efficiency uptake must be dizzying!

9. The size of theropod skulls implies they had big brains, perhaps made possible by a high metabolic rate. But brain size is hard to interpret because many reptiles don't fill their cranium with brain. Traces on theropod skulls suggest that blood vessels serving the brain touched the skull, indicating their brain did fill the cranium, but that's hardly conclusive. And there are cheaper ways to build a big brain than being hot-blooded; there's no necessary connection.

10. The evidence for all this is preserved in the rocks as 'isotopic signatures'. For those who want to know more, I'll be so bold as to recommend an article I wrote for *Nature* on the subject: 'Reading the Book of Death', *Nature* (July, 2007).

Chapter 9

1. According to Michael Gazzaniga, in his book *The Social Brain*, his mentor Roger Sperry had returned from a conference at the Vatican, and remarked that the Pope had said, in essence if not these exact words, 'that the scientists could have the brain and the Church would have the mind'.

2. I am adopting a dualistic vocabulary – assuming a distinction between mind and brain – even though I do not think there is a distinction, partly to emphasise how deeply embedded within the language such dualism is, and partly because it reflects the explanatory challenge. If the mind and the brain are the same thing, then we must explain why they do not seem to be. Merely to say 'It's all an illusion!' is not good enough; what is the molecular basis of the illusion?

3. The original Deep Blue lost a series to Kasparov in 1996, despite winning a single game. An updated version, known unofficially as Deeper Blue, beat Kasparov in a series in 1997, but Kasparov said he sometimes saw 'deep intelligence and creativity' in the computer's

moves, and accused IBM of cheating. On the other hand, if a coterie of computer programmers can defeat a genius at chess, the implications are nearly as bad – genius by committee.

4. For anyone interested in more of these strange cases I must recommend V. S. Ramachandran's fascinating books, deeply grounded as they are in neurology and evolution.

5. Oscillations are rhythmic changes in the electrical activity of single neurons, and if enough act together in concert they can be picked up on an EEG. When a neuron fires it depolarises, which is to say, the membrane charge partly dissipates, as ions like calcium and sodium rush into the cell. If neuronal firing is random and sporadic, it is hard to pick up anything organised on an EEG; but if large numbers of neurons across the brain depolarise and then repolarise in rhythmic waves, the effects can be picked up as brain waves on an EEG. Oscillations in the region of 40 Hz mean that many neurons are firing in synchrony, roughly every 25 milliseconds.

6. Synapses are tiny sealed gaps between neurons, which physically break the passage of a nerve impulse (which is to say, they interrupt firing). Chemicals called neurotransmitters are released when an impulse arrives at a synapse. These diffuse across the gap, then bind to receptors on the 'post-synaptic' neuron, either activating or inhibiting it, or inducing longer-term changes that strengthen or weaken synapses. The formation of new synapses or alteration of existing synapses is involved in the formation of memories and learning, though much remains to be discovered about the detailed mechanisms.

7. There's even some tantalising evidence that consciousness is composed of 'still frames' in much the same way as a movie. These frames can vary in duration from a few tens of milliseconds to a hundred or more. Shortening or lengthening these frames, under the influence of emotions, for example, may account for why time seems to slow down or speed up under different circumstances. So time would slow down fivefold if frames were built every 20 milliseconds rather than every 100 milliseconds: we would see an arm wielding a knife move in slow motion.

8. Dark matter is the stuff of consciousness, and described as 'Dust' in the 'His Dark Materials' trilogy of novelist Philip Pullman, I presume in homage to James's 'mind-dust'.

Chapter 10

1. Technically bacteria and plants possess 'metacaspase' enzymes, not true caspases; but the metacaspases are plainly the evolutionary precursors of the caspases found in animals, and serve many similar purposes. For simplicity, I shall refer to them all as caspases. For more detail, see my article 'Origins of Death', *Nature* (May, 2008).

2. Enzyme cascades are important in cells because they amplify a small starting signal. Imagine that a single enzyme is activated, which in turn activates ten target enzymes, each of which activates another ten targets. We now have a hundred active enzymes. If each activates another ten targets we have a thousand, then ten thousand, and so on. It only takes six steps in such a cascade to activate a million executioners that tear apart the cell.

3. There were other reasons, of course, that set the eukaryotes on the path to multicellularity, while the bacteria never really developed beyond colonies, in particular the tendency of eukaryotic cells to grow larger and accumulate genes. The reasons behind this development are a major theme of my earlier book *Power, Sex, Suicide: Mitochondria and the Meaning of Life*.

4. I don't know whether I'm harbouring any *ApoE4* variants, but from the spectrum of diseases that run in my family I wouldn't be entirely surprised to find I had at least one. That's why I'd rather not know. I ought to get down the gym.

5. Kirkwood called his theory the 'disposable soma theory', recollecting Weismann's terms of reference. The body is subservient to the germ-line, Kirkwood and Weismann say in unison; and the Pacific salmon is an exemplary case.

6. There may be some unexpected drawbacks. One man who subjected himself to a strict calorie restriction regime unexpectedly broke his leg

after a minor fall. He had developed serious osteoporosis, and his doctor, rightly, warned him off the dietary regime.

7. For those who really want to know how a molecule can be 'sensitive' to the presence or absence of nutrients, SIRT-1 binds to and is activated by the 'spent' form of a respiratory coenzyme called NAD, which only builds up in the cell when substrates such as glucose are depleted. TOR is 'redox sensitive' which means that its activity differs according to the oxidation state of the cell, which again reflects the availability of nutrients.

8. We noted another possible trade-off earlier: cancer versus degenerative disease. Mice with an extra gene for SIRT-1 show signs of better health, but don't live longer. Instead they mostly die of cancer, an unfortunate trade-off.

9. Other postulated 'clocks', like the length of telomeres (the caps on the end of chromosomes that shorten with every cell division) don't correlate at all with lifespan across species. While correlation doesn't prove causality, it's a good starting point. Lack of correlation more or less disproves causality. Whether the telomeres protect against cancer, by blocking limitless cell division, is a moot point, but they certainly don't determine lifespan.

10. A few examples should clarify my meaning. I'm not talking about the kind of acute inflammation we see in an inflamed cut. Atherosclerosis entails a chronic inflammatory reaction to material accumulating in arterial plaques, and ongoing inflammation exacerbates the process. Alzheimer's disease is driven by a persistent inflammatory reaction to amyloid plaques in the brain; age-related macular degeneration involves inflammation of the retinal membranes, leading to the in-growth of new blood vessels and blindness. I could go on – diabetes, cancer, arthritis, multiple sclerosis, and more. Chronic mild inflammation is the common denominator. Smoking provokes such diseases largely by exacerbating inflammation. Conversely, blocking TOR induces mild immunosuppression, as we've seen, and this might help to dampen inflammation.

11. The idea of a glycolytic switch goes back to Otto Warburg in the 1940s, but has recently been verified. As a general rule only cells that can do without their mitochondria turn cancerous. The greatest culprits are stem cells, which have little dependence on mitochondria and are often implicated in tumorogenesis. Otherwise skin cells, lung cells, white blood cells, all are relatively independent of mitochondria, and all are associated with tumours.

12. As Gustavo Barja notes, the fact that evolution can extend lifespan by a whole order of magnitude implies that a very significant prolongation of human lifespan is quite feasible, merely exacting.

LIST OF ILLUSTRATIONS

3.4 The ancient mineral structure of the oxygen-evolving complex – four manganese atoms linked by oxygen in a lattice, with a calcium atom nearby, as revealed by X-ray crystallography. Redrawn from: Yano J, *et al*. Where Water Is Oxidised to Dioxygen: Structure of the Photosynthetic Mn4Ca Cluster *Science* 314: 821; 2006.

4.1 Differences between prokaryotic cells like bacteria, and complex eukaryotic cells.

4.2 Conventional tree of life.

4.3 Tree of life based on ribosomal RNA.

4.4 The 'ring of life' redrawn from: Rivera MC, Lake JA. 'The ring of life provides evidence for a genome fusion origin of eukaryotes', *Nature* 431: 152–155; 2004.

4.5 Bacterial cells living within other bacterial cells. Courtesy of Carol von Dohlen, Utah State University.

4.6 Structure of nuclear membrane. Reproduced with permission from Bill Martin: Archaebacteria and the origin of the eukaryotic nucleus. *Current Opinion in Microbiology* 8: 630–637; 2005.

5.1 The spread of new beneficial mutations in sexual (top) versus asexual (bottom) organisms.

6.1 The structure of skeletal muscle, showing the characteristic striations and bands. Courtesy of Professor Roger Craig, University of Massachusetts.

6.2 Myosin, in the exquisite watercolour of David Goodsell. Courtesy of Dr David Goodsell, Scripps Research Institute, San Diego.

6.3 Actin filaments derived from the slime mould *Physarum polycephalum*, decorated with myosin 'arrowheads' from rabbit muscle. Reproduced with permission from Hugh Huxley: Nachmias VT, Huxley HE, Kessler D. Electron microscope observations on actomyosin and actin preparations from Physarum polycephalum, and on their interaction with heavy meromyosin subfragment I from muscle myosin. *J Mol Biol* 1970: 50; 83–90.

6.4 Actin cytoskeleton in a cartilage cell from a cow, labelled with the fluorescent dye phalloidin-FITC. Courtesy of Dr Mark Kerrigan, University of Westminster.

ACKNOWLEDGEMENTS

This has not been a lonely book. For much of the time I have had my two young sons, Eneko and Hugo, about the house, and their presence, though not always conducive to concentration, has lent meaning and pleasure to every word. My wife, Dr Ana Hidalgo, has discussed every theme, every idea, every word, often bringing new perspectives to bear, and cutting out junk with unsentimental scissors. I have grown to depend on the trueness of her judgement, both scientific and literary, and where once I used to argue, only to reflect and concede that I was wrong, I am now swift to take her advice. This book would have been far worse without her, and many of its merits are hers.

And then there have been many stimulating conversations by email from around the world, at all times of the day and night, the thoughts and perspectives of specialists in the diverse fields I have tried to cover. While I have always followed my own path, I have greatly appreciated their generosity and expertise. In particular, I thank Professors Bill Martin, at Heinrich Heine University, Düsseldorf, John Allen, at Queen Mary, University of London, and Mike Russell, now at the NASA Jet Propulsion Laboratory, Caltech. Powerful and original thinkers all, I am indebted for their time and encouragement, robust criticism, and inspiring love of science. If ever my own enthusiasm waned, an email or meeting with them was always galvanic.

But they were not alone, and I'm very grateful to a number of other researchers for clarifying their ideas, and for reading and commenting on chapters. Each chapter has benefited from the constructive criticism of at least two specialists in the field. I thank (in alphabetical order): Professor Gustavo Barja, Complutense University, Madrid, Professor Bob Blankenship, Washington University, Professor Shelley Copley, University of Colorado, Dr Joel Dacks, University of Alberta, Professor Derek Denton, University of Melbourne, Professor Paul Falkowski, Rutgers University, New Jersey, Professor Hugh Huxley, Brandeis University, Massachusetts, Professor Marcel Klaasen, Netherlands Institute of Ecology, Professor Christof Koch, Caltech, Dr Eugene Koonin, National Institutes of Health, Maryland, Professor Pavel Koteja, Jagiellonian University, Poland, Professor Michael Land, University of Sussex, Professor Björn Merker, University of Uppsala, Professor Salvador Moncada, University College London, Professor José Musacchio, New York University, Professor Sally Otto, University of British Colombia, Professor Frank Seebacher, University of Sydney, Dr Lee Sweetlove, University of Oxford, Dr Jon Turney, and Professor Peter Ward, University of Washington. Any remaining errors are my own.

I thank my family, both here and in Spain, for their love and support. My father in particular has taken time from writing his own books on history to read and comment on almost all the chapters, overcoming his dislike of molecules. I'm very grateful too to a curiously shrinking circle of friends, who on my third book still read chapters and comment. Thanks especially to Mike Carter, whose generosity is second to none, even in the most testing times; to Andrew Philips, for stimulating discussions and kind comments; and to Paul Asbury for climbs, walks and talks. More generally, I thank Professor Barry Fuller, for squash and regular discussions of science over a pint; Professor Colin Green, for his sweeping interests, and for believing in me when I needed it; Dr Ian Ackland-Snow, for unbounded enthusiasm, and reminding me how lucky I am at regular intervals; Dr John Emsley, for many stimulating conversations on science writing over the years, and for helping to get me started as a writer; Professor Erich and Andrea Gnaiger, whose hospitality knows no bounds; and Mr Devani and Mr Adams, inspiring teachers, who many years ago sowed the seeds of my lifelong love of biology and chemistry.

Finally a note to my editors at Profile and WW Norton, Andrew Franklin and Angela von der Lippe, who have been unflagging in their support and belief in this book from the beginning; to Eddie Mizzi, for his good taste and eclectic knowledge in copy-editing; and to Caroline Dawnay, my agent at UA, who is always ready with a kind word of encouragement, without ever losing her bracing perspective.

London 2009

BIBLIOGRAPHY

I have written *Life Ascending* mostly from primary sources, but the books highlighted below were all sources of insight and enjoyment for me. I don't agree with everything in all of them, but that is part of the pleasure of a good book. Most are written for the general reader and are original scientific works in the tradition of *The Origin of Species* itself. Alphabetical order.

David Beerling *The Emerald Planet* (OUP, 2006). A splendid book on the impact of plants on the history of our planet. Colourful extended vignettes of evolution.

Susan Blackmore *Conversations on Consciousness* (OUP, 2005). An admirably level-headed attempt to canvas and make sense of the conflicting views of leading scientists and philosophers on consciousness.

Jacob Bronowski *The Ascent of Man* (Little, Brown, 1974). The book of the TV series. Magnificent.

Graham Cairns-Smith *Seven Clues to the Origin of Life* (CUP, 1982). Scientifically an old book now, but unequalled for its Sherlock Holmesian examination of the clues at the heart of life's chemistry.

— *Evolving the Mind* (CUP, 1998). An excellent and unusual defence of quantum theories of consciousness from a determinedly individual thinker.

Francis Crick *Life Itself* (Simon & Schuster, 1981). An indefensible
hypothesis defended by one of the finest scientific minds of the twentieth
century. Dated, but still worth a read.

Antonio Damasio *The Feeling of What Happens* (Vintage Books, 2000).
One of several books by Damasio, all well worth a read. Poetic and
intense; at its best on neurology, and weakest on evolution.

Charles Darwin *The Origin of Species* (Penguin, 1985). Among the most
important books ever written.

Paul Davies *The Origin of Life*. (Penguin, 2006). A fine book, if perhaps
prone to see too many obstacles.

Richard Dawkins *Climbing Mount Improbable* (Viking, 1996). One of
surprisingly few popular books to deal explicitly with the evolution of
complex traits like eyes, in Dawkins' inimitable style.

— *The Selfish Gene* (OUP, 1976). One of the defining books of our age. A
must-read.

Daniel Dennett *Consciousness Explained* (Little, Brown, 1991). A
controversial classic that can't be ignored.

Derek Denton *The Primordial Emotions* (OUP, 2005). An eloquent book
with an important thesis, all the better for its excellent discussions of the
work of other researchers and philosophers.

Christian de Duve *Singularities* (CUP, 2005). A densely argued book,
encapsulating much of de Duve's clear thinking. Not for the chemically
faint-hearted, but the distillations of a fine mind.

— *Life Evolving* (OUP, 2002). An elegiac book on the biochemical
evolution of life and our place in the universe. It reads like a swan song,
but in fact precedes *Singularities*.

Gerald Edelman *Wider than the Sky* (Penguin, 2004). A slim but dense
volume, and a good introduction to Edelman's important ideas.

Richard Fortey *Trilobite!* (HarperCollins 2000). A delight, as are all
Fortey's books, and especially good on the evolution of the trilobites'
crystal eye.

Tibor Ganti *The Principles of Life* (OUP, 2003). An original take on the
fundamental make-up of life, if lacking in real-world biochemistry. I
reluctantly did not discuss his ideas.

Steven Jay Gould *The Structure of Evolutionary Theory* (Harvard University Press, 2002). A serious work of scholarship, but with flashes of Gould's customary panache; a goldmine of interesting material.

Franklin Harold *The Way of the Cell* (OUP, 2001). A wonderful, insightful book on the cell, full of the poetry of deep science. Not always easy but well worth the effort.

Steve Jones *The Language of the Genes* (Flamingo, 2000). Wonderful introduction to the gene; effervescent.

— *Almost like a Whale* (Doubleday, 1999). Darwin updated with panache and deep learning.

Horace Freeland Judson *The Eighth Day of Creation* (Cold Spring Harbor Press, 1996). A seminal tome: interviews with the pioneers of molecular biology, at the dawn of an era.

Tom Kirkwood *Time of Our Lives* (Weidenfeld & Nicolson, 1999). One of the best introductions to the biology of ageing, by a pioneer in the field; a humane work of science.

Andre Klarsfeld and Frederic Revah *The Biology of Death* (Cornell University Press, 2004). A thoughtful examination of aging and death, from the point of view of the cell. A rare meld of evolution and medicine.

Andrew Knoll *Life on a Young Planet* (Princeton University Press, 2003). A fine, accessible view of early evolution by one of the leading protagonists. Wise counsel.

Christof Koch *The Quest for Consciousness* (Roberts & Co, 2004). A textbook, and not an easy read, but full of flair and insight. A fine and balanced mind enquiring into the mind.

Marek Kohn *A Reason for Everything* (Faber and Faber, 2004). Short biographies of five British evolutionists, full of eloquence and insight. Very good on the evolution of sex.

Michael Land and Dan-Eric Nilsson *Animal Eyes* (OUP, Oxford, 2002). A textbook on animal optics, accessibly written and conveying well the exuberant ingenuity of evolution.

Nick Lane *Power, Sex, Suicide* (OUP, 2005). My own exploration of cell biology and the evolution of complexity, from the point of view of cellular energy.

— *Oxygen* (OUP, 2002). A history of life on earth with the gas that made complex life possible, oxygen, taking centre-stage.

Primo Levi *The Periodic Table* (Penguin, 1988). Not really a science book at all, but brimming with poetry, humanity and science. A master literary craftsman.

Lynn Margulis and Dorion Sagan *Microcosmos* (University of California Press, 1997). A fine introduction to the grand scale of the microcosm from one of the iconic figures in biology, no stranger to controversy.

Bill Martin and Miklós Müller *Origin of Mitochondria and Hydrogenosomes* (Springer, 2007). A scholarly multi-author text on a seminal event in evolution: disparate views of the eukaryotic cell.

John Maynard Smith and Eörs Szathmáry *The Origins of Life* (OUP, 1999). A popular reworking of their classic text *The Major Transitions in Evolution* (WH Freeman/Spektrum, 1995). A major work by great intellects.

John Maynard Smith *Did Darwin get it Right?* (Penguin, 1988). Accessible essays, some on the evolution of sex, by the late guru of evolutionary biologists.

Oliver Morton *Eating the Sun* (Fourth Estate, 2007). A gem of a book, combining a novelist's insight into people and place with a deep feeling for molecules and planets. Especially good on the carbon crisis.

Andrew Parker *In the Blink of an Eye* (Free Press, 2003). Much to savour, despite a rather blinkered view.

Vilayanur Ramachandran *The Emerging Mind* (Profile, 2003). Ramachandran's 2003 Reith lectures, essentially his earlier book *Phantoms in the Brain* in compressed form. An original and imaginative mind at play.

Mark Ridley *Mendel's Demon* (Weidenfeld & Nicolson, 2000). An intellectual treat, lacing its profound material on the evolution of complexity with grave wit.

Matt Ridley *The Red Queen* (Penguin, 1993). A seminal book on the evolution of sex and sexual behaviour; insightful and full of Ridley's trademark flair.

— *Francis Crick* (HarperPress, 2006). A first-rate biography of one of the most interesting figures in twentieth century science. Nicely nuanced, without ever jeopardising his love of the subject.

Steven Rose *The Making of Memory* (Vintage 2003). A fine book on the neural events underlying memory, at its best on the social framework of science.

Ian Stewart and Jack Cohen *Figments of Reality* (CUP, 1997). An eclectic, witty and erudite look at consciousness. Entertaining and opinionated.

Peter Ward *Out of Thin Air* (Joseph Henry Press, 2006). A serious and original hypothesis about why dinosaurs were dominant for so long. Accessible and compelling.

Carl Zimmer *Parasite Rex* (The Free Press, 2000). A superb book on the importance of parasites, with an excellent section on the role of parasites in the evolution of sex.

PRIMARY SOURCES

Francis Crick once complained that 'There is no form of prose more difficult to understand and more tedious to read than the average scientific paper'. He had a point – but he did use the word 'average'. At their best, scientific papers can be pure distillations of meaning, and capable of exercising as much pull on the mind as a work of art. I have tried to limit my list here to papers of that calibre: this is not an exhaustive list, but a selection of the papers that most influenced my thinking in writing this book. I've also added a few general reviews as entrées into the literature. Alphabetical order by chapter.

Introduction

Russell R.J., Gerike U., Danson M.J., Hough D.W., Taylor G.L. Structural adaptations of the cold-active citrate synthase from an Antarctic bacterium. *Structure* 6: 351-61; 1998.

Chapter 1: The Origin of Life

Fyfe W. S. The water inventory of the Earth: fluids and tectonics. Geological Society, London, Special Publications 78: 1–7; 1994.

Holm N. G., *et al*. Alkaline fluid circulation in ultramafic rocks and formation of nucleotide constituents: a hypothesis. *Geochemical Transactions* 7:7; 2006.

Huber C., Wächtershäuser G. Peptides by activation of amino acids with CO on (Ni,Fe)S surfaces: implications for the origin of life. *Science* 281: 670–72; 1998.

Kelley D. S., Karson J. A., Fruh-Green G. L. *et al*. A serpentinite-hosted ecosystem: the Lost City hydrothermal field. *Science* 307: 1428–34; 2005.

Martin W., Baross J., Kelley D., Russell M. J. Hydrothermal vents and the origin of life. *Nature Reviews in Microbiology* 6: 805–14; 2008.

Martin W., Russell M. J. On the origin of biochemistry at an alkaline hydrothermal vent. *Philosophical Transactions of the Royal Society of London B* 362: 1887–925; 2007.

Morowitz H., Smith E. Energy flow and the organisation of life. *Complexity* 13: 51–9; 2007.

Proskurowski G., *et al*. Abiogenic hydrocarbon production at Lost City hydrothermal field. *Science* 319: 604–7; 2008.

Russell M. J., Martin W. The rocky roots of the acetyl CoA pathway. *Trends in Biochemical Sciences* 29: 358–63; 2004.

Russell M. First Life. *American Scientist* 94: 32–9; 2006.

Smith E., Morowitz H. J. Universality in intermediary metabolism. *Proceedings of the National Academy of Sciences USA* 101: 13168–73; 2004.

Wächtershäuser G. From volcanic origins of chemoautotrophic life to bacteria, archaea and eukarya. *Philosophical Transactions of the Royal Society of London B* 361: 1787–806; 2006.

Chapter 2: DNA

Baaske P., *et al*. Extreme accumulation of nucleotides in simulated hydrothermal pore systems. *Proceedings of the National Academy of Sciences USA* 104: 9346–51; 2007.

Copley S. D., Smith E., Morowitz H. J. A mechanism for the association of amino acids with their codons and the origin of the genetic code. PNAS 102: 4442-7 2005.

Crick F. H. C. The origin of the genetic code. *Journal of Molecular Biology* 38: 367–79; 1968.

De Duve C. The onset of selection. *Nature* 433: 581–2; 2005.

Freeland S. J., Hurst L. D. The genetic code is one in a million. *Journal of Molecular Evolution* 47: 238–48; 1998.

Gilbert W. The RNA world. *Nature* 319: 618; 1986.

Hayes B. The invention of the genetic code. *American Scientist* 86: 8–14; 1998.

Koonin E. V., Martin W. On the origin of genomes and cells within inorganic compartments. *Trends in Genetics* 21: 647–54; 2005.

Leipe D., Aravind L., Koonin E.V. Did DNA replication evolve twice independently? *Nucleic Acids Research* 27: 3389–401; 1999.

Martin W., Russell M. J. On the origins of cells: a hypothesis for the evolutionary transitions from abiotic geochemistry to chemoautotrophic prokaryotes, and from prokaryotes to nucleated cells. *Philosophical Transactions of the Royal Society of London B.* 358: 59–83; 2003.

Taylor F. J. R., Coates D. The code within the codons. *Biosystems* 22: 177–87; 1989.

Watson J. D., Crick F. H. C. A structure for deoxyribose nucleic acid. *Nature* 171: 737–8; 1953.

Chapter 3: Photosynthesis

Allen J. F., Martin W. Out of thin air. *Nature* 445: 610–12; 2007.

Allen J. F. A redox switch hypothesis for the origin of two light reactions in photosynthesis. *FEBS Letters* 579: 963–68; 2005.

Dalton R. Squaring up over ancient life. *Nature* 417: 782–4; 2002.

LIFE ASCENDING

Ferreira K. N. *et al*. Architecture of the photosynthetic oxygen-evolving center. *Science* 303: 1831–8; 2004.

Mauzerall D. Evolution of porphyrins – life as a cosmic imperative. *Clinics in Dermatology* 16: 195–201; 1998.

Olson J. M., Blankenship R. E. Thinking about photosynthesis. *Photosynthesis Research* 80: 373–86; 2004.

Russell M. J., Allen J. F., Milner-White E. J. Inorganic complexes enabled the onset of life and oxygenic photosynthesis. In *Energy from the Sun: 14th International Congress on Photosynthesis*, Allen J. F., Gantt E., Golbeck J. H., Osmond B. (editors). *Springer* 1193–8; 2008.

Sadekar S., Raymond J., Blankenship R. E. Conservation of distantly related membrane proteins: photosynthetic reaction centers share a common structural core. *Molecular Biology and Evolution* 23: 2001–7; 2006.

Sauer K., Yachandra V. K. A possible evolutionary origin for the Mn_4 cluster of the photosynthetic water oxidation complex from natural MnO_2 precipitates in the early ocean. *Proceedings of the National Academy of Sciences USA* 99: 8631–6; 2002.

Walker D. A. The Z-scheme – Down Hill all the way. *Trends in Plant Sciences* 7: 183–5; 2002.

Yano J., *et al*. Where water is oxidised to dioxygen: structure of the photosynthetic Mn_4Ca cluster. *Science* 314: 821–5; 2006.

Chapter 4: The Complex Cell

Cox C. J., *et al*. The archaebacterial origin of eukaryotes. *Proceedings of the National Academy of Sciences USA* 105: 20356–61; 2008.

Embley M. T , Martin W. Eukaryotic evolution, changes and challenges. *Nature* 440: 623–30; 2006.

Javaeux E. J. The early eukaryotic fossil record. In: *Origins and Evolution of Eukaryotic Endomembranes and Cytoskeleton* (Ed. Gáspár Jékely); Landes Bioscience 2006.

Koonin E. V. The origin of introns and their role in eukaryogenesis: a
    compromise solution to the introns-early versus introns-late debate?
    Biology Direct 1: 22; 2006.

Lane N. Mitochondria: key to complexity. In: *Origin of Mitochondria and
    Hydrogenosomes* (Eds Martin W, Müller M); Springer, 2007.

Martin W., Koonin E. V. Introns and the origin of nucleus-cytosol
    compartmentalisation. *Nature* 440: 41–5; 2006.

Martin W., Müller M. The hydrogen hypothesis for the first eukaryote.
    Nature 392: 37–41; 1998.

Pisani D, Cotton J. A., McInerney J. O. Supertrees disentangle the
    chimerical origin of eukaryotic genomes. *Molecular Biology and Evolution*
    24: 1752–60; 2007.

Sagan L. On the origin of mitosing cells. *Journal of Theoretical Biology* 14:
    255–74; 1967.

Simonson A. B., *et al.* Decoding the genomic tree of life. *Proceedings of the
    National Academy of Sciences USA* 102: 6608–13; 2005.

Taft R. J., Pheasant M., Mattick J. S. The relationship between non-protein-
    coding DNA and eukaryotic complexity. *BioEssays* 29: 288–99; 2007.

Vellai T., Vida G. The difference between prokaryotic and eukaryotic cells.
    Proceedings of the Royal Society of London B 266: 1571–7; 1999.

Chapter 5: Sex

Burt A. Sex, recombination, and the efficacy of selection: was Weismann
    right? *Evolution* 54: 337–51; 2000.

Butlin R. The costs and benefits of sex: new insights from old asexual
    lineages. *Nature Reviews in Genetics* 3: 311–17; 2002.

Cavalier-Smith T. Origins of the machinery of recombination and sex.
    Heredity 88: 125–41; 2002.

Dacks J., Roger A. J. The first sexual lineage and the relevance of
    facultative sex. *Journal of Molecular Evolution* 48: 779–83; 1999.

Felsenstein J. The evolutionary advantage of recombination. *Genetics* 78:
    737–56; 1974.

Hamilton W. D., Axelrod R., Tanese R. Sexual reproduction as an adaptation to resist parasites. *Proceedings of the National Academy of Sciences USA* 87: 3566–73; 1990.

Howard R. S., Lively C. V. Parasitism, mutation accumulation and the maintenance of sex. *Nature* 367: 554–7; 1994.

Keightley P. D., Otto S. P. Interference among deleterious mutations favours sex and recombination in finite populations. *Nature* 443: 89–92; 2006.

Kondrashov A. Deleterious mutations and the evolution of sexual recombination. *Nature* 336: 435–40; 1988.

Otto S. P., Nuismer S. L. Species interactions and the evolution of sex. *Science* 304: 1018–20; 2004.

Szollosi G. J., Derenyi I., Vellai T. The maintenance of sex in bacteria is ensured by its potential to reload genes. *Genetics* 174: 2173–80; 2006.

Chapter 6: Movement

Amos L. A., van den Ent F., Lowe J. Structural/functional homology between the bacterial and eukaryotic cytoskeletons. *Current Opinion in Cell Biology* 16: 24–31; 2004.

Frixione E. Recurring views on the structure and function of the cytoskeleton: a 300 year epic. *Cell Motility and the Cytoskeleton* 46: 73–94; 2000.

Huxley H. E., Hanson J. Changes in the cross striations of muscle during contraction and stretch and their structural interpretation. *Nature* 173: 973–1954.

Huxley H. E. A personal view of muscle and motility mechanisms. *Annual Review of Physiology* 58: 1–19; 1996.

Mitchison T. J. Evolution of a dynamic cytoskeleton. *Philosophical Transactions of the Royal Society of London B* 349: 299–304; 1995.

Nachmias V. T., Huxley H., Kessler D. Electron microscope observations on actomyosin and actin preparations from *Physarum polycephalum*, and on their interaction with heavy meromyosin subfragment I from muscle myosin. *Journal of Molecular Biology* 50: 83–90; 1970.

OOta S., Saitou N. Phylogenetic relationship of muscle tissues deduced from superimposition of gene trees. *Molecular Biology and Evolution* 16: 856–67; 1999.

Piccolino M. Animal electricity and the birth of electrophysiology: The legacy of Luigi Galvani. *Brain Research Bulletin* 46: 381–407; 1998.

Richards T. A., Cavalier-Smith T. Myosin domain evolution and the primary divergence of eukaryotes. *Nature* 436: 1113–18; 2005.

Swank D. M., Vishnudas V. K., Maughan D. W. An exceptionally fast actomyosin reaction powers insect flight muscle. *Proceedings of the National Academy of Sciences USA* 103: 17543–7; 2006.

Wagner P. J., Kosnik M. A., Lidgard S. Abundance distributions imply elevated complexity of post-paleozoic marine ecosystems. *Science* 314: 1289–92; 2006.

Chapter 7: Sight

Addadi L., Weiner S. Control and Design Principles in Biological Mineralisation. *Angew Chem Int Ed Engl* 3: 153–69; 1992.

Aizenberg J., *et al.* Calcitic microlenses as part of the photoreceptor system in brittlestars. *Nature* 412: 819–22; 2001.

Arendt D., *et al.* Ciliary photoreceptors with a vertebrate-type opsin in an invertebrate brain. *Science* 306: 869–71; 2004.

Deininger W., Fuhrmann M., Hegemann P. Opsin evolution: out of wild green yonder? *Trends in Genetics* 16: 158–9; 2000.

Gehring W. J. Historical perspective on the development and evolution of eyes and photoreceptors. *International Journal of Developmental Biology* 48: 707–17; 2004.

Gehring W. J. New perspectives on eye development and the evolution of eyes and photoreceptors. *Journal of Heredity* 96: 171–84; 2005.

Nilsson D. E., Pelger S. A pessimistic estimate of the time required for an eye to evolve. *Proceedings of the Royal Society of London B* 256: 53–8; 1994.

Panda S., *et al.* Illumination of the melanopsin signaling pathway. *Science* 307: 600–604; 2005.

Piatigorsky J. Seeing the light: the role of inherited developmental cascades in the origins of vertebrate lenses and their crystallins. *Heredity* 96: 275–77; 2006.

Shi Y., Yokoyama S. Molecular analysis of the evolutionary significance of ultraviolet vision in vertebrates. *Proceedings of the National Academy of Sciences USA* 100: 8308–13; 2003.

Van Dover C. L., *et al.* A novel eye in 'eyeless' shrimp from hydrothermal vents on the Mid-Atlantic Ridge. *Nature* 337: 458–60; 1989.

White S. N., *et al.* Ambient light emission from hydrothermal vents on the Mid-Atlantic Ridge. *Geophysical Research Letters* 29: 341–4; 2000.

Chapter 8: Hot Blood

Burness G. P., Diamond J., Flannery T. Dinosaurs, dragons, and dwarfs: the evolution of maximal body size. *Proceedings of the National Academy of Sciences USA* 98: 14518–23; 2001.

Hayes J. P., Garland J. The evolution of endothermy: testing the aerobic capacity model. *Evolution* 49: 836–47; 1995.

Hulbert A. J., Else P. L. Membranes and the setting of energy demand. *Journal of Experimental Biology* 208: 1593–99; 2005.

Kirkland J. I., *et al.* A primitive therizinosauroid dinosaur from the Early Cretaceous of Utah. *Nature* 435: 84–7; 2005.

Klaassen M., Nolet B. A. Stoichiometry of endothermy: shifting the quest from nitrogen to carbon. *Ecology Letters* 11: 1–8; 2008.

Lane N. Reading the book of death. *Nature* 448: 122–5; 2007.

O'Connor P. M., Claessens L. P. A. M. Basic avian pulmonary design and flow-through ventilation in non-avian theropod dinosaurs. *Nature* 436: 253–6; 2005.

Organ C. L., *et al.* Molecular phylogenetics of Mastodon and Tyrannosaurus rex. *Science* 320: 499; 2008.

Prum R. O., Brush A. H. The evolutionary origin and diversification of feathers. *Quarterly Review of Biology* 77: 261–95; 2002.

Sawyer R. H., Knapp L. W. Avian skin development and the evolutionary origin of feathers. *Journal of Experimental Zoology* 298B: 57–72; 2003.

Seebacher F. Dinosaur body temperatures: the occurrence of endothermy and ectothermy. *Paleobiology* 29: 105–22; 2003.

Walter I., Seebacher F. Molecular mechanisms underlying the development of endothermy in birds (*Gallus gallus*): a new role of PGC-1α? *American Journal of Physiology Regul Integr Comp Physiol* 293: R2315–22, 2007.

Chapter 9: Consciousness

Churchland P. How do neurons know? *Daedalus* Winter 2004; 42–50.

Crick F., Koch C. A framework for consciousness. *Nature Neuroscience* 6: 119–26; 2003.

Denton D. A., *et al.* The role of primordial emotions in the evolutionary origin of consciousness. *Consciousness and Cognition* 18: 500–514; 2009.

Edelman G., Gally J. A. Degeneracy and complexity in biological systems. *Proceedings of the National Academy of Sciences USA* 98: 13763–68; 2001.

Edelman G. Consciousness: the remembered present. *Annals of the New York Academy of Sciences* 929: 111–22; 2001.

Gil M., De Marco R. J., Menzel R. Learning reward expectations in honeybees. *Learning and Memory* 14: 49–96; 2007.

Koch C., Greenfield S. How does consciousness happen? *Scientific American* October 2007; 76–83.

Lane N. Medical constraints on the quantum mind. *Journal of the Royal Society of Medicine* 93: 571–5; 2000.

Merker B. Consciousness without a cerebral cortex: A challenge for neuroscience and medicine. *Behavioral and Brain Sciences* 30: 63–134; 2007.

Musacchio J. M. The ineffability of qualia and the word-anchoring problem. *Language Sciences* 27: 403–35; 2005.

Searle J. How to study consciousness scientifically. *Philosophical Transactions of the Royal Society of London B* 353: 1935–42; 1998.

Singer W. Consciousness and the binding problem. *Annals of the New York Academy of Sciences* 929: 123–46; 2001.

Chapter 10: Death

Almeida A., Almeida J., Bolaños J. P., Moncada S. Different responses of astrocytes and neurons to nitric oxide: the role of glycolytically-generated ATP in astrocyte protection. *Proceedings of the National Academy of Sciences USA* 98: 15294–99; 2001.

Barja G. Mitochondrial oxygen consumption and reactive oxygen species production are independently modulated: implications for aging studies. *Rejuvenation Research* 10: 215–24; 2007.

Bauer *et al*. Resveratrol improves health and survival of mice on a high-calorie diet. *Nature* 444: 280–81; 2006.

Bidle K. D., Falkowski P. G. Cell death in planktonic, photosynthetic microorganisms. *Nature Reviews in Microbiology* 2: 643–55; 2004.

Blagosklonny M. V. An anti-aging drug today: from senescence-promoting genes to anti-aging pill. *Drug Discovery Today* 12: 218–24; 2007.

Bonawitz N. D., *et al*. Reduced TOR signaling extends chronological life span via increased respiration and upregulation of mitochondrial gene expression. *Cell Metabolism* 5: 265–77; 2007.

Garber K. A mid-life crisis for aging theory. *Nature* 26: 371–4; 2008.

Hunter P. Is eternal youth scientifically plausible? *EMBO Reports* 8: 18–20; 2007.

Kirkwood T. Understanding the odd science of aging. *Cell* 120: 437–47; 2005.

Lane N. A unifying view of aging and disease: the double-agent theory. *Journal of Theoretical Biology* 225: 531–40; 2003.

Lane N. Origins of death. *Nature* 453: 583–5; 2008.

Tanaka M., *et al*. Mitochondrial genotype associated with longevity. *Lancet* 351: 185–6; 1998.

INDEX

dinucleotide 50–51
disease
  age-related 271, 272–3, 275, 281, 283, 284
  cardiovascular 271, 273, 284
  degenerative 269
  infectious 270
disequilibrium
  chemical 14, 17, 23
disposable soma theory 304n
DNA (deoxyribonucleic acid) 4, 5, 10, 11, 14, 18, 33, 34–6, 39–43, 49, 51, 56–8, 90, 100, 153, 243, 279
  base pairing 35–6, 43, 50–51
  chloroplast 73
  degeneracy 44, 47
  differences with RNA 58
  doublet code 41, 48
  fingerprinting 37
  'letters' 35–9, 42, 98
  non-coding 91, 96, 112–13, 203
  overlapping codes 42, 44
  polymerase 290n
  reading frames 44, 290n
  replication 55, 290n, 291n
  sequences 36–39
  triplet code 41, 46–8, 50, 96
  universal code 41, 45, 55
dog 185
domains (of life) 100, 289n
Doolittle, Ford 88
double helix 34, 35, 39
Double Helix, The 40
dragonfly (giant) 64–5, 145
Drosophila (see fruit fly)
dualism (Cartesian) 233, 302n
Duncan, Isadora 295n

E

Earth 1, 8, 9, 12
Eating the Sun 291n
ecology 133

ecosystem 145
  complexity of 145–7
  marine 146
Edelman, Gerald 245–8, 255
EEG (electroencephalogram) 243, 303n
egg cell (see oocyte)
Egyptians, ancient 255
Einstein, Albert 205
electricity 149
  animal 150, 158
  potential difference 153, 195
electromagnetic radiation 252, 291n
electron 13, 68–72, 81, 83, 85–6, 87, 151, 252, 288n
electron microscopy (see microscopy)
electron-transport chain 80–81, 83, 84
Eliot, T. S. 244
Else, Paul 301n
embryological development 218–19, 301n
  growth axes 301n
Emeishan traps 225
emotion 235, 239, 244, 249
  versus feeling 244–5
endoplasmic reticulum 113
Entobdella soleae 299n
  mitochondrial lens 299n
enzyme 6, 27–8, 49, 53, 55–6, 58, 100, 185, 190–91
  cascade 304n
  detoxifying 278
epilepsy 234, 240
  epileptic seizures 234, 240
erosion 65
Escherichia coli (E. coli) 103
eubacteria (see bacteria)
eukaryote 90, 95, 99, 101–2, 140, 164
  common ancestor 133, 139, 170
eukaryotic cell 4, 89, 92, 93, 98, 100, 105, 106, 112, 116, 140–41, 267, 289n
  as chimera 102–5
  competition with bacteria 116